MILITARY CULTURE AND EDUCATION

Military Culture and Education

Edited by
DOUGLAS HIGBEE
University of South Carolina, Aiken, USA

Routledge
Taylor & Francis Group
LONDON AND NEW YORK

First published 2010 by Ashgate Publishing

Published 2016 by Routledge
2 Park Square, Milton Park, Abingdon, Oxfordshire OX14 4RN
711 Third Avenue, New York, NY 10017, USA

First issued in paperback 2016

Routledge is an imprint of the Taylor & Francis Group, an informa business

British Library Cataloguing in Publication Data
Military culture and education.
1. Military education--United States. 2. Soldiers--Education, Non-military--United States. 3. College teachers--United States. 4. Civil-military relations--United States.
I. Higbee, Douglas.
355'.00711'73-dc22

Library of Congress Cataloging-in-Publication Data
Higbee, Douglas, 1970-
Military culture and education / by Douglas Higbee.
p. cm.
Includes bibliographical references and index.
ISBN 978-1-4094-0757-7 (hardback)
1. Military education--United States. 2. Military socialization--United States. 3. United States--Armed Forces--Officers--Training of. 4. Military education. 5. Military socialization. 6. Armed Forces--Officers--Training of. I. Title.

U408.H54 2010
355.50973--dc22

2010020076

ISBN 13: 978-1-138-25679-8 (pbk)
ISBN 13: 978-1-4094-0757-7 (hbk)

Contents

Notes on Contributors

Thomas Bruscino is Assistant Professor of Military History at the US Army School of Advanced Military Studies, Fort Leavenworth, Kansas. He is the author of *A Nation Forged in War: How World War II Taught Americans to Get Along* (University of Tennessee Press, 2010) and his articles, review essays, and opinion pieces have appeared in the *Claremont Review of Books*, *Military Review*, *War and Society*, *Journal of America's Military Past*, *Doublethink*, *San Luis Valley Historian*, and *Reviews in American History*.

Carol Burke is Professor of English, Faculty Associate in Anthropology, and Director of Religious Studies at the University of California, Irvine. She embedded with an Army Human Terrain System team in northern Iraq in December 2008 and January 2009 and has published three articles based on that experience. Her last book, *Camp All-American, Hanoi Jane, and the High-and-Tight*, examined contemporary military culture and its resistance to change. Burke's current project looks at the ways in which the military struggles to understand and adapt to the cultures in the countries it occupies.

Bradley L. Carter is Associate Professor of Military History, US Army's Command and General Staff College, Fort Leavenworth, Kansas. Currently assigned to the CGSC satellite at Fort Gordon, Georgia, he also serves as a faculty Staff Group Advisor. He earned a PhD in American Studies at the University of Kansas, is a graduate of the Seabury-Western Theological Seminary, and is a former administrative dean at Saint Paul School of Theology. An active member of the Society for Military History, Carter presented "Mars turns to Minerva: The Military, Social Science, and War in the 21st Century" at Cambridge University's Center for International Studies 2009 conference.

D. Alexis Hart was formerly an officer in the United States Navy and is currently Assistant Professor of Rhetoric and Writing at the Virginia Military Institute. She is co-editor of an upcoming special issue of *Kairos* on rhetoric, technology, and the military, and is a co-author of an upcoming book, *Publishing a Journal of Undergraduate Research*.

Douglas Higbee earned his PhD from the University of California, Irvine and is Assistant Professor of English at the University of South Carolina, Aiken. He has published articles on war poetry, including Siegfried Sassoon and Ivor Gurney, as well as on British veterans' organizations.

Daniel J. Hughes retired from the Air War College in July 2009 after 18 years on the faculty. He earned a PhD in history from the University of North Carolina in 1979 and served as an Army officer in the Vietnam War. He previously worked at the infantry School at Fort Benning and the Combined Arms Center at Fort Leavenworth, and taught in professional military education programs at all levels from pre-commissioning to senior service schools, as well as at colleges in Georgia, Kansas and Alabama. His publications include *The King's Finest: A Social Profile of Prussia's General Officers 1871-1914* and (as translator and editor) *Moltke on the Art of War: Selected Writings.*

Andrew H. Myers is Associate Professor of American Studies at the University of South Carolina, Upstate and a lieutenant colonel in the United States Army Retired Reserve. He received the Combat Infantryman Badge and Bronze Star Medal for the service in Iraq he recounts in this book. Trained as a social historian, he has spent more than sixteen years studying the army as a domestic institution. He is author of *Black, White and Olive Drab: Racial Integration at Fort Jackson, South Carolina, and the Civil Rights Movement.*

Huw Osborne is Assistant Professor in the Department of English and a Writing Centre Coordinator at the Royal Military College of Canada. His most recent research concerns the life and writing of the Welsh author Rhys Davies. His monograph on Davies was published by the University of Wales Press in 2009.

Edward F. Palm, a Vietnam veteran and a retired Marine officer turned academic, served as the Marine Officer Instructor with the Naval ROTC unit at the University of California, Berkeley and taught English at the US Naval Academy. He has published extensively on military culture and literature. Since leaving the military, he has been a tenured full professor at Glenville State College (WV) and has held dean appointments at Maryville University of St. Louis and Olympic College (Bremerton, WA). He is currently Professor of English with the online division of Strayer University.

Karen P. Peirce is Graduate Writing Coordinator at North Dakota State University. Previously, she was Assistant Professor of English at the United States Military Academy. She has published articles in *CEA Forum*, *WPA Journal*, and *Rhetoric Review*, and she is Managing Editor of *FYHC: The Journal of First-Year Honors Composition.*

Jeffrey S. Sychterz is Assistant Professor of English at Fayetteville State University in North Carolina. He graduated from the US Naval Academy and served as a naval officer before receiving his PhD from the University of Illinois, Urbana-Champaign. He has published essays and given several papers on war poetry and related topics.

Andrea Trocha-Van Nort, an Oklahoma native, lived in France and taught English, literature, economics, and management there in private schools, private universities, and four different public universities over fifteen years. She returned to the United States in 2007 to teach English literature at the United States Air Force Academy. Her research is concentrated in sixteenth- and seventeenth-century theater studies.

Major David C. Wood is a career US Army officer. His assignment to the United States Military Academy will end summer 2010 after teaching Composition, Literature, and Advanced Composition for three years. He will be returning to the Army's deploying combat brigades. During his time at West Point, he has presented papers at conferences in the areas of American Civil War reconciliation and multi-genre writing. He is writing an article on Writing Across the Curriculum between the Math and Composition program at West Point prior to his departure.

Preface

This volume aims to transform an imaginary community into an actual one, or at least to hasten that process. This imaginary community is the vast but diffuse collection of university-level teachers and scholars with feet in both academics and the military. We know who we are, and we know there are many others like us out there. And we know that people who work in only one of these cultures, the academy or the military, are still rather unaware or uninformed about how these two cultures intersect. We hope to be a part of the process by which we begin to build stronger networks that link our work and experiences together as well as to continue to examine, and perhaps bridge, the gap between the academy and the military.

I would first like to acknowledge the work and perseverance of the contributors to this volume. Next, thanks to Dr Suzanne Ozment, Executive-Vice Chancellor for Academic Affairs at the University of South Carolina, Aiken, for supporting my work on this book with a research grant. I also thank Ashgate Publishing for its interest in the volume and its editors for their expert and gracious support. Lastly, I thank my wife, Sharon, for her love and patience.

Douglas Higbee
March 17, 2010

Introduction
Intersections

Douglas Higbee

A fundamental component of the United States' system of government places the president, an elected civilian, as the commander in chief of the armed forces. This arrangement helps hold together what in many ways are opposing forces: on the one hand, the president is the highest official in a political system that functions chiefly via rhetorical power; on the other hand, the president is the head of a military whose chief function is to exercise sheer physical force. For a democracy, having the president serve (together with the assistance of the civilian secretary of defense) as the leader of the military has the clear advantage of limiting the autonomy of the military, for an independent military can lead to militarism or worse. At the same time, having a commander-in-chief whose stock in trade is verbal suasion carries with it the possibility of the military being used for partisan reasons. Juggling these two roles well is likely why George Washington, Abraham Lincoln, and Franklin Roosevelt are commonly considered to the United States' greatest presidents; the problems of Lyndon Johnson's and George W. Bush's presidencies are likewise attributed to their inability to adequately balance both roles.

Nevertheless, this tension between the military and civilian society as embodied in the president has been primarily a healthy one. For a key factor in American political, economic, and social life, and those of other liberal democracies, has been the perennial existence of a tension between competing claims and ideals; debate and dissent are the engines for growth and discovery in a democratic society. Again, problems in this arrangement tend to occur when this civil-military relation tilts out of balance—when, for instance, the last president to have had any meaningful military experience was George H.W. Bush, who served as a naval aviator in World War II. This is not necessarily to impugn specific political leaders, nor am I trying to argue that people with military service necessarily make better commanders-in-chief. This is just to point out that since the framers of our Constitution structured this basic overlap or tension between the political and military spheres, this arrangement has usually been accompanied by chief executives who have some acquaintance with the military. It is probably no coincidence that this political and cultural tension's recent exacerbation coincides roughly with the end of the military draft in the 1970s.

Academe and the Military: A Healthy Tension?

If many of our recent leaders have not had significant military experience, a concurrent development in this growing distance between the political and martial worlds has been the rather conflicted relationship between the military and the academy. While in the post-World War II era initiatives such as the GI Bill, which introduced millions of veterans to higher education, helped strengthen the relationship between the academy and the military, this rough partnership was weakened in the wake of the Vietnam War, especially as service deferments were granted for college and graduate school attendance. This practice, coupled with the controversial nature of the war itself, tended to divide the Vietnam generation into academics and members of the military. Together with such trends, developments within the academy and the military themselves further separated these two institutions. In the post-draft era, the military has became increasingly professionalized and insular, while in the academy the sway of theoretical and discipline-specific practices and vocabulary, especially in the humanities and social sciences, has likewise contributed to the distance between scholars and primary currents in American society, including the military.

This recent historical division between academe and the military has reinforced several key cultural differences, broadly speaking. Alex Vernon, who as a West Point graduate, Gulf War veteran, and current Professor of English, has lived for quite some time on the academic-military fault line. On the one hand, he notes the perennial and often reflexive tendency of academics to decry "the evils of the patriarchal military-industrial complex" (2002). On the other hand, in his powerful memoir *Most Succinctly Bred* Vernon describes his experience as a cadet as inimical to his burgeoning vocation as a writer; he was the only literature major in a class of 1,000 (2006: 79). Through years of living betwixt and between the two cultures, he has compiled a working list of their most salient polarities: while academe tends to value tolerance, dialogue, and creativity, and strives to nurture multiple points of view, the military tends to value authority, utilitarianism, and physical force (2002). William Corley, also a former officer and current English professor, describes the cultural differences between the military and academe in terms of the age-old figures of the sword and pen: as he writes, "academe educates future policy makers; the military executes that policy" (2006). While Vernon tends to see these differences as largely solidified, Corley and others, as well as many contributors to this volume, have emphasized that such differences are more nuanced than plain, more partial than total.[1]

To be sure, current developments in the academic-military relationship have been controversial. In particular, social scientists have been concerned about the direct involvement of anthropologists and others in military operations such as the Human Terrain System, which uses such academics as cultural advisors to combat

1 For a variety of compelling views on the relation between military culture and the humanities, see Farrell 2004, Niday and Harrington 2007, Samet 2007, and B. Hanley 2008.

units in the ongoing wars in Iraq and Afghanistan. Academics crossing over into military advisement, some argue, compromises scholarly objectivity as they deploy the tools and knowledge of social science in the service of military intelligence.[2] This rift within academe has its counterpart in the military itself. For instance, Gen. David Petraeus, himself a Princeton PhD, has argued for the relevance of advanced academic study for officers. Graduate school, he argues, creates "flexible, adaptive, creative thinkers" essential to fighting counter-insurgency wars effectively (2007). In contrast, Ralph Peters, a retired Army officer and writer, in using Hamlet as his chief example argues that "too much formal education clouds a senior officer's judgment, inhibits his instincts, and slows his decision making" (2007). These issues as outlined in the Petraeus-Peters debate articulate a major concern of this collection, especially in the chapters written by humanities teachers working at military institutions.

It's important to note, in other words, that both the military and academe are quite diverse institutions—it's difficult to just say "the academy" or "the military" without being more specific as to what kind of institution of higher education, what military branch and unit you mean. And the diversity within both institutions creates points of overlap between them: for instance, a supply or public relations officer in the military has more in common with a manager at United Parcel Service or a White House spokesperson than with an Army Ranger or fighter pilot. Likewise, an adjunct instructor at a community college has little in common with a professor conducting research at a prestigious university, but his or her commitment to students is often fairly analogous to a junior officer's devotion to his or her troops. Lastly, within the university system, programs devoted to scientific research, business and other more vocational types of education are more congruent with the ethos of large corporations and military contractors than with the liberal arts sectors of academe.

Another similarity is that both institutions are structured hierarchically—like the military, the academe is not a democracy but a permeable, and largely meritocratic, class system. At academic conferences attendees wear ID badges that specify their name and school—which is tantamount to unit or even rank, since different schools have different statuses within the profession. The leaders in academe, in today's increasingly corporatized educational environment, are not "grunts" or even junior officers; they are CEO-type administrators who have often never taught in a college classroom—never been "in the trenches"—or at least have not for a considerable time. This academic hierarchy itself has been regulated and negotiated via a comprehensive system of education and training: the more you know and learn, the higher you can ascend in the organizational ladder. Of course, the military tends to emphasize technical (which often has a physical component) training, while academe prefers the kinds of intellectual and

---

2 For more on the continuing controversy over the Human Terrain System, see R. Gonzalez 2007, Connable 2009, and Glenn 2009, as well as Chapter 2 by Carol Burke in this volume.

moral learning emphasized in education. But probably the most significant point of overlap between academe and the military, at least for this collection, is that many academics and soldiers tend to be idealists—both groups conceive their jobs as a vocation and a service to the larger society; both groups' activities are (somewhat) insulated from market forces, and both groups are paid significantly less than those with comparable expertise in the private sector.

The Chapters in this Collection

My prime motivation in putting together this collection, then, comes from a strong sense that there is significant overlap—and significant tensions within this overlap—between academe and military culture with which we haven't yet come to grips. In part, this motivation is personal: my father attended the US Military Academy and served as a career Army officer, including a tour in Vietnam and a three-year stint teaching at West Point. Though I did not follow in his military footsteps, in my time in academe I have not left behind an interest in military culture, especially in its complex relation to the rest of society. In addition to my own family background, I sensed that there were many others like myself with a personal and/or professional foot in both cultures, and I wanted to provide a coherent forum for their stories and opinions.

But while all the pertinent aspects of the intersection between higher education and military culture need parsing, it's not possible to address all the points of overlap and tension between these two institutions in a single book. Our collection, for its part, emphasizes the interactions as seen in the humanities and social sciences, from the point of view of people currently teaching in military schools and civilian universities. Our collective contention is that we need to be aware of, to even welcome, the tensions between these two cultures—in the intersection between scholars and the military outside the classroom, in military academies, and in professional military schools.

Part I of this collection, "Intersections In and Out of the Field," presents three chapters from writers with extensive personal experience in the various ways academe and military culture intersect, beginning with Ed Palm's piece. His deep investment in both the military and academe comes through in a compelling retrospective of his career as an officer and a humanities scholar. Finding that academics, particularly those in the humanities, are often treated dismissively within the nearly "monastic" military culture of Annapolis, he ends up arguing for the augmentation of Reserve Officer Training Corps (ROTC) programs in order to address the continuing divisions between military and civilian life. Carol Burke, a professor of English and Folklore and a longtime commentator on military culture, examines the military's use of anthropologists and other social scientists in its Human Terrain System program. Based on her experience as an embedded reporter with an HTS team in Iraq, Burke argues that the program's inadequacies can be addressed with better preparation of civilian academics for the exigencies of work in war zones and with augmented funding for social science research. Lastly, Andrew Myers, a professor of American Studies, served as leader of a military-transition team in northern Iraq from 2006 to 2007. His chapter describes the challenges facing his unit, including the deteriorating security conditions and the poorly-trained state of the Iraqi Army units. In all, however, Dr Myers emphasizes the applicability of his academic training and perspective to his role as military advisor.

Part II, "Military Academies and Humanistic Inquiry," offers five chapters by English and writing professors at military academies in the United States and Canada. Former Navy officer Alexis Hart, now teaching at the Virginia Military Institute, argues that the ancient Greek pedagogy of Isocrates—which stresses both rhetorical competence and civic engagement—fosters her students' eagerness to become active citizens while offering an antidote to the individualism and consumerism especially endemic among young Americans. Hart finds quite a strong and felicitous overlap in her writing classroom between academic and military culture. Jeff Sychterz, a professor of English and former Navy officer, likewise finds midshipmen at the Naval Academy more willing to challenge American social norms than students he has encountered at civilian universities. Perhaps because his tenure teaching at Annapolis was more recent, he offers a more optimistic assessment of the confluence between academic and military culture than Palm. Karen Peirce, a civilian professor, and Dave Wood, a career Army officer, offer a dialogue on the challenges and opportunities involved in teaching writing at West Point. While noting the institutional and pedagogical constraints present at that military academy, both assert the value of both subjects in the formation of capable and self-aware military leaders.

A civilian professor of English at the US Air Force Academy, in her classroom Andrea Trocha-Van Nort negotiates the potential disjunction between the humanities and officer training by foregrounding the ethical quandaries presented in classic works of literature, including *Beowulf*, *Hamlet*, and *A Farewell to Arms*. She finds that emphasizing the ethical dimension both stimulates her student-cadets' critical thinking skills while providing them with a means with which to make the complex transition from civilian to officer. Huw Osborne, a civilian professor at Canada's premier military academy, the Royal Military College, offers a cogent analysis of cadet culture in terms of the tension between officially-sanctioned homosociality and the homophobic attitudes of many cadets. While realistic about the challenges in using literature to foster more tolerant attitudes toward homosexuality, Osborne is optimistic about the prospects for continued progress at an institution still struggling to redefine itself within contemporary Canadian society.

In Part III, "Teaching in Professional Military Schools," three professors of military history discuss the challenges and opportunities attendant to teaching in graduate-level military schools.[3] Tom Bruscino, a civilian professor of military history at the School for Advanced Military Studies at Fort Leavenworth, argues that the military is still struggling to infuse its education of senior military officers in strategy and national policy with academic rigor, or what Bruscino calls "expertise." While granting that this dual mission will always be accompanied by a certain tension, his prescription for a curriculum more focused on specific domains of applicability and on inculcating critical thinking skills in his students promises to strike a workable balance. Also written from the point of view of a civilian academic, Dan Hughes paints a less sanguine view of education in professional military schools. Surveying his eighteen years at the Air War College, he describes the near-total absence of academic values and standards in a cautionary series of episodes geared toward would-be professors at that institution. Some readers may very well disagree with Hughes' findings, while others may point out that other professional military schools do not fit his analysis of the AWC, but I think all readers will appreciate Hughes' forthright appraisal and dry humor. Lastly, military historian Bradley Carter's experience as a civilian academic at the Army's Command and General Staff College at Fort Gordon, Georgia illuminates both the cultural differences between academe and the military that are at the heart of this volume and provides insight into the ways in which scholarly history can advance the goals of an institution tasked with the advanced education of field-grade officer.

3 For a good overview and analysis of professional military education, see Stiehm 2002.

PART I
Intersections In and Out of the Field

Chapter 1
Real Officers Don't Teach Keats: The Naval Academy, ROTC, and Military Spiritualism

Edward F. Palm

The ghost of James Webb haunts the halls of the English Department at the United States Naval Academy. This is quite an accomplishment, considering that Webb isn't dead yet. A Naval Academy graduate of the class of '68, a former Marine officer, a highly decorated Vietnam veteran, a Georgetown Law School graduate, a best-selling and critically acclaimed novelist, a former Secretary of the Navy, and now a US senator—Webb always was an overachiever. But one of his unsung accomplishments was to position himself precisely at the intersection of the military and academic cultures at the Naval Academy, an institution uniquely illustrative of how those cultures come into conflict and where they find common ground.

Following the success of his first novel, *Fields of Fire* (1978), Webb wangled an appointment to his alma mater as a Visiting Assistant Professor. He taught there in the English Department from January 15 to June 15, 1979, breaking what was to have been an 18-month contract in order to pursue "other opportunities."[1] Anyone familiar with the pressures and problems of academic staffing would assume that there was, and perhaps still is, some bitterness over Webb's early departure. It would also be understandable for those who had committed themselves heart and soul to the academic profession to resent having an apparent dilettante thrust upon them, especially one whose growing fame and fortune gave him the freedom to move on at will. But Webb's ghostly presence in Sampson Hall, the longstanding home of the English Department, is not simply a matter of jealousy or resentment.

A House Divided

Ever since its inception in 1845, the Naval Academy has rested upon a myth: Academy founder George Bancroft so revered education that he hired a cadre of civilian professors to do it right. And, for the most part, thus it has remained. Only a handful of officers are permanently assigned to the Naval Academy. The great

1 Erickson, J., 2009. "James Webb and USNA." [Email] Message to Edward Palm. Sent May 28, 2009.

majority only pass through on two- or three-year tours of duty.[2] The permanent faculty is predominantly civilian. In principle, this arrangement should make Navy the most progressive and enlightened of the academies—an enviable alliance between Athens and Sparta. In practice, it has resulted in a "house divided"—an uneasy and unequal marriage of convenience between two cultures, neither of which very much likes or trusts the other. Truth be told, the Naval Academy's civilian faculty is not the hallmark of any great concern for education; rather, it is the mark of the Navy's indifference.

Up until recently, the other academies have gone too far in the other direction, selecting virtually all their faculty from among their active duty ranks. This has long struck their critics as academically incestuous and, in the early nineties, began to impress Congress as wasteful of operationally trained officers in that era of defense cutbacks. Accordingly, West Point and the Air Force Academy both were ordered to look toward the Naval Academy model and to start incorporating civilian academics into their ranks. Navy's supporters in Congress, however, still cannot or will not acknowledge the downside of civilianizing a service academy. Air Force and Army, by having officers fully involved in academics as well as leadership, send a clear signal to their cadets that both services do value education. (Both of these academies, moreover, have kept civilians in the minority, and neither one grants tenure to civilian faculty.) Navy, on the other hand, in relegating the business of education largely to civilians, leaves its midshipmen (a traditional title Navy insists on applying to women as well as men) wondering what the Navy really does value and expect from them.

The Naval Academy has always been quick to point out that a significant number of officers are assigned to the academic divisions. The current Web site boasts that the academy's "600-member faculty is an integrated group of officers and civilians in nearly equal numbers" (United States Naval Academy 2009). In my experience, however, the roughly 300 officers counted toward that total includes company officers and others assigned to leadership or administrative billets. These officers do not teach in any formal sense and can hardly be considered "faculty." Only in recent years have a handful of permanent military faculty members been appointed.[3] But, again, the great majority of officers assigned to the Naval

2 I taught in the Naval Academy's English Department from 1990-1993. I was a military faculty member, a Marine major, at the time; even though I had a PhD, college teaching experience, and publications, the Marine Corps would not allow me to extend my assignment beyond three years, and I was told that there was "no way" the department could hire me to a civilian appointment following my impending retirement.

3 I flatter myself that a brash article I wrote for *The Chronicle of Higher Education* in 1996 played some small role in bringing about this reform. See Palm 1996. As of this writing, 33 Permanent Military Professors (PMPs) are assigned to the Naval Academy. Including Weapons and Systems units, 13 are assigned to engineering departments, ten to sciences, eight to liberal arts disciplines, and two to leadership (Halbig, M., "RE: Permanent Military Professors." [Email] Message to Edward Palm. Sent June 8, 2009).

Academy, in either instructional or leadership roles, merely pass through on their way back to the operating forces.

This is in marked contrast to the model adopted by the United States Military Academy at West Point. Its faculty still consists primarily of military officers, most of whom have earned master's degrees at their service's expense and who pass through on two- or three-year tours of duty. As directed by Congress, both Army and Air Force do now employ a limited number of civilian academics on fixed-term, and even some renewable, contracts, but the only permanent, or "tenured," appointments at these academies are held by the military faculty. West Point also has an elite group of 28 military-academic elders, the term for which is "Professor, USMA." These are generally full colonels who, by special federal statute, may remain on active duty at West Point until age 64½ and who retire as brigadier generals. One member of this group serves as the academic dean, and the rest chair the academic departments.[4] While the Naval Academy, in recent years, has appointed 33 permanent military professors, academic leadership remains largely in the hands of a civilian dean and a civilian faculty.[5]

Palm: Professor or Major, USMC?

From its point of view, the Naval Academy selects officers to be military role models first and teachers second. An officer's primary responsibility is to recruit for his or her warfare specialty and to the keep the "mids" focused on what is ultimately important—the prized commission that awaits them upon graduation. This means that, for the military faculty, teaching takes a back seat to meetings, social gatherings, and military training. (A fellow military faculty member once reminded me of the academy-correct attitude and set of priorities: "I'm a Marine first, a pilot second, and a professor third," he declared.) On all occasions, in or out of the classroom, officers are expected to show themselves to be approachable,

4 Dalton, D.R. "Academic Rank of the Permanent Military Faculty." [Email] Message to Edward Palm. Sent June 9, 2009.

5 The Naval Academy has opted not to adopt the elite West Point full professor model. Its permanent military professors may remain at the Naval Academy beyond the normal career limits imposed on line officers of the same rank—24 years for a commander, 30 for a Navy captain—the equivalent of an Army or Navy colonel (Halbig, M. "RE: Permanent Military Professors." [Email] Message to Edward Palm. Sent June 8, 2009). Most PMPs have begun as assistant professors and must earn academic promotion in the same manner as their civilian colleagues. Recently, the Navy has given its PMPs the option of applying to remain on active duty until age 65, but retention past 28 or 30 years is not guaranteed (Halbig, follow-up, June 8, 2009). Also, an informed source at the Naval Academy tells me that there is some concern about PMPs losing touch with what is happening in the fleet and that regular short-duration, temporary assignments with the operating forces are being considered. This would suggest that the Navy expects a PMP to be a military role model first and an academic second.

personable, and personally impressive. The goal is not just to be liked, but like Arthur Miller's Willy Loman, to be "well-liked." Among the more irreverent officers and mids, this is known as the "shiny shoes, empty head syndrome."

A case in point is what it took to get me, as a Marine major, to Annapolis in the first place. Soon after reporting, I learned that the military hierarchy had been largely indifferent to my PhD, academic publications, and college teaching experience. They were seriously concerned about my atypical career pattern and the reputation I had established as an outspoken liberal loose cannon. (They were apparently indifferent to my enlisted service in Vietnam.) The civilian chair of my department, I later learned, had taken no position. Two colonels for whom I had worked, one a Naval Academy alumnus, had to vouch for my military professionalism before the academy would accept me.

Holding a PhD in English, I soon realized, was not a mark in my favor. Real Navy and Marine officers don't teach Keats.[6] All the services have certain prescribed career patterns designed to keep officers focused on their operational roles, but given its mission, the Navy tends to discourage idiosyncratic interests more than the other services. Graduate schooling followed by a pay-back teaching tour at the academy is not considered to be career-enhancing. This is especially the case if an officer commits him- or herself to a liberal arts field instead of engineering, math, or one of the hard sciences. But regardless of teaching field, an officer who wants to keep getting promoted in the Navy understands that he or she has to get back to the fleet. The "integration" the academy's Web site touts, therefore, is more rhetorical than real. The civilian faculty predominate, and the Navy has allowed them to set up essentially a university departmental structure within a military command. The problem is that the values, ideals, and expectations on both sides are decidedly different.

Athens Versus Sparta

The most glaring difference involves the time-honored academic tradition of shared governance. Military organizations are hierarchical. There is a well-defined "chain of command," and while this structure admits more collegial interaction and collaboration than the average civilian realizes, a military command is essentially an authoritarian organization. There is one individual in charge, the commander, who is clearly empowered to make all the decisions and is ultimately responsible for whatever the command does or fails to do. The commander, moreover, is not popularly elected but rather appointed by an officer above him or her in the chain of command.

6 This was the original title of my 1996 *Chronicle* article on the academy and one I would have preferred to keep. I was able to use it in a revision published in *Roofbeam*, a now defunct literary magazine. See Palm 1998.

The Naval Academy is unique among the service academies in that a largely self-governing civilian academic organization exists within a hierarchical, authoritarian military command. The academic divisions, which are comparable to colleges at civilian universities, are almost exclusively headed by senior military officers. But the department chairs are mostly civilian academics elected by their colleagues to fixed terms. In my experience, moreover, military division heads generally display the wisdom to leave the running of the academic program largely to the faculty.

This was certainly the case in the division in which I taught, the Division of English and History (which has since been subsumed into a larger organization, the Division of Humanities and Social Sciences). The division was, and as of this writing still is, headed by a Marine colonel, but my department chair was a civilian faculty member whom the faculty had elected to a four-year term. Likewise, tenure and promotion decisions are made very much the way they would be made at a university—through a process of peer review, with tenure and promotion committees elected from among the tenured faculty. While senior military faculty members are usually invited to serve on these committees, they are always in the minority. The academic peer groups at USNA are overwhelmingly civilian. As of this writing, for instance, the English Department has 25 civilian and eight military faculty members, only one of whom is a permanent military professor (United States Naval Academy 2009).

While earning tenure in USNA's English Department is hardly a Darwinian publish-or-perish struggle, peer-reviewed scholarship is the expectation. Other departments have set the same standard and, along with English, have negotiated a three-credit release from the 12-credit teaching load specified as the contract requirement—the academy ostensibly being committed, first and foremost, to teaching. This release has been a constant source of tension between the faculty and the administration.

But if shared governance has always been important to the English faculty at the Naval Academy, the right to select and hire their colleagues is one the civilian faculty long ago appropriated and jealously guards. They form the hiring committees, they interview at the annual Modern Language Association convention, they invite a few leading candidates to campus, and they make the final hiring decisions—just as it is typically done in English departments at civilian colleges and universities throughout the nation.[7]

7 Their hiring committees usually include a token officer, but in keeping with the composition of the department at large, the majority of the membership is always civilian. One justification for this practice is that, up until recently, officers come and go, but the civilian faculty has to live with a new civilian hire for more than two or three years, possibly for an entire career.

Enter James Webb's Ghost

It is unclear just who secured Webb's appointment to the Naval Academy in 1979, but it certainly was not initiated by the English Department faculty. And here is where the military and academic cultures do not just intersect; they collide.

From the military's standpoint, Webb would seem to have been ideally qualified to teach at the academy. Not only was Webb an alumnus: he had gone on to become a Marine officer, a highly decorated combat veteran, a law-school graduate, and a best-selling and critically-acclaimed novelist. He was not an academic in the traditionalist sense, but the military hierarchy clearly considered him to be academic enough. The Naval Academy, after all, is just that—a service academy and not a university.

But Webb was hardly academic enough to suit the civilian faculty. Their criteria included a PhD, scholarly promise in a traditional or contemporary literary specialty, and a demonstrated commitment to the academic profession as a career. For the military hierarchy, however, what Webb lacked in traditional academic credentials, he more than made up for as a military role model. Webb's assignment, in sum, challenged the assumptions of the civilian faculty, some of whom feared that the soldier-poet, and not the professional academic, would win out as the preferred model for a Naval Academy faculty member. The situation was aggravated only a few years later, when a Navy officer with a PhD was assigned to finish out his career teaching in the English Department. This officer applied for a civilian position in the department but was not even granted an interview. Upon his retirement, he wrote to the superintendent charging the English Department with being anti-military. As a consequence, the administration appointed a military officer to chair the department. The administration also intervened in one of the department's searches, insisting that the department hire the candidate the administration preferred—a PhD who also happened to be a Naval Reserve officer. Only two or three years later, the department resumed their practice of electing chairs from among their civilian colleagues. In all fairness, however, the civilian chair is generally assisted by an associate chair that he or she selects from among the military faculty.[8]

"The Hall" Versus "the Yard"

Aside from threatening the civilian faculty's hold on hiring, Webb's short stint at USNA points up a longstanding, and still unresolved, debate about the nature and proper role of education in an institution committed to preparing young men and women for leadership in the Navy and Marine Corps. Academy officials are quick to tout education as "Job Number One," the first priority at the Naval

8 This could perhaps be justified by the nature of the military-civil relation in America. Still, I find it highly ironic in a service academy to find the military answerable to civilian academics.

Academy. But as any member of the departments commonly referred to as the "bull departments"—English, History, and Political Science—will privately admit, the Navy hierarchy understands and values education only in the narrow scientific and technical sense. At his retirement ceremony in Annapolis, an outgoing Chief of Naval Operations laid it on the line when he urged the Brigade of Midshipmen to major in math or science lest they find themselves "wallflowers in the ballroom of progress" (Anderson 2008).

The entire academy program, moreover, is designed to keep midshipmen overburdened in pursuit of what one superintendent, in my hearing, termed "balanced excellence." The mids routinely carry upwards of 20 credit hours every semester and must participate in a number of military and extracurricular activities. Sleep deprivation and constant stress are both intended results. When I was there, new faculty members were urged not to take it personally and to be lenient when mids fall asleep in class. The point of it all, as I heard another superintendent say, was to force the mids "to compartmentalize their lives the way an officer in combat must compartmentalize his." The emphasis has long been on mission accomplishment and, only in recent years (and largely as the result of a few well-publicized scandals), has the Naval Academy begun to pay serious attention to the societal and ethical implications of the mission.

What the academy has never understood, or intentionally ignores, is that liberal education is antithetical to compartmentalization. It depends upon having "world enough and time" to draw parallels and make connections between disciplines.[9] The Naval Academy, it has always seemed to me, encourages midshipmen to view education, whether technical or liberal, as only the means to a pragmatic end.

While today's mids are free to major in a full range of traditional academic majors, including the "bull subjects" of English, history, and political science, they must also take a core engineering curriculum that includes calculus, physics, and electrical engineering—hence, the requirement to carry upwards of 20 credit hours every semester.

This insistence on making every graduate an engineer is an article of faith shared by all the federal service academies and reflects a presumption born of military hubris. It is telling that the services use the term "doctrine" to denote the prescribed operating procedures and the steps to take in the face of various anticipated tactical contingencies. Military manuals and handbooks tend to present war as a kind of calculus problem, one a well-prepared officer can solve through a correct application of the appropriate "doctrine." The military has always been loathe to admit the role of chance and circumstance in a battle's outcome. Military training and education programs, ironically, tend to give short shrift to military history, which is replete with accounts of the best laid plans gone awry and victory snatched from the jaws of defeat through spontaneous feats of daring and improvisation the enemy never anticipated.

9 For a detailed defense of literature as an integral part of military education, see Palm 1986b.

Military history notwithstanding, true believers still cling to this article of faith: "Everyone knows they have to be engineers." An elderly woman, the wife of a retired Naval Academy alumnus, told me this upon learning that I was teaching English at the academy. She went on to date the ruination of the academy from the introduction of liberal arts majors. The Navy, of course, can make a stronger case than the other services. It needs nuclear engineers to run its submarines and its nuclear-powered surface ships as well as its complex weapons systems. But the great majority of Naval officers, including academy graduates, serve in other capacities. "Poly Sci and fly!" many a midshipman hoping to secure the requisite GPA for flight school has been heard to say.

Still, there is a method to the madness at Annapolis. Regardless of academic major or ultimate ambition, all the mids are hoisted on the academy's hallowed version of Neptune's trident. The academic program constitutes only one prong of that trident. The other two are athletics and military performance (which includes leadership).[10] This pursuit of "balanced excellence" is not for "pencil-necked geeks," as I heard a superintendent remind a faculty assembly. In all fairness to those out of sympathy with "geeks" or nerds, the most effective officers, and the ones destined for high command, do tend to have excellent people skills and can stand up to a great deal of pressure. Toward this end, the entire program—academics, athletics, and military performance—is intended to weed out loners and to induce stress. The justification is that anyone who cannot handle whatever the academy can throw at him or her cannot be trusted with the responsibility for the lives of our sailors and Marines or the safe operation of a multimillion dollar ship, plane, or weapons system. To a civilian faculty member who must fight to keep his mids awake, however, the phrase "balanced excellence" has an Orwellian ring, suggesting more of enforced mediocrity than actual excellence.

All three of our federal service academies have always been committed to the premise that adversity builds character—which is one of the principal tenets of military spiritualism. Military spiritualists believe that the profession of arms is a sacred trust requiring its followers to lead a largely unexamined life. They subscribe to certain articles of faith, such as "the more we sweat in peace, the less we bleed in war." "Everyone knows they all have to be engineers" is another one. The Naval Academy lexicon likewise includes floating signifiers such as "patriotism" and "professionalism"—the latter functioning as a kind of shibboleth. Everyone at the academy aspires to or espouses "professionalism," but no one actually defines it. The term merely connotes the approved attitude of unquestioning respect for

10 Since I left the academy, leadership and athletics have been included in a larger weighted category called "Aptitude for Commissioning," which counts for 17.68 percent of a midshipman's class standing. Academics and professional course grades account for 64.48 percent (Halbig, M., "RE: Permanent Military Professors." [Email] Message to Edward Palm. Sent June 17, 2009). It is telling that professional and academic courses are lumped together here. The company officer, in concert with the officers teaching professional courses, can still determine a midshipman's future.

authority and devotion to duty. A civilian faculty committed to the unfettered search for truth and to intellectual independence often find this level of discourse to be embarrassing, so much so that they deride it—but generally only in private among themselves.

This is not to say that all military officers, or even the great majority, are military spiritualists. But their influence is still felt in the conviction, shared by military training programs in general, that artificially induced stress surmounted today can condition the recipient to stand up to the stresses of combat tomorrow. That influence is also felt in the enduring conviction that an engineering curriculum works best to prepare officers to improvise when the unexpected happens and a vital system fails. Naval officers, more than their counterparts in the other services, do indeed need to understand and use complex technology. But, as many a professor in the "bull departments" would argue, the curriculum that has long been associated with making people imaginative, flexible, resourceful, and responsible is more liberal than technical.

To their credit, the civilian faculty at Annapolis work hard at developing the traditional liberal arts skills, values, and habits of mind in all the mids. The problem is that the academy has a way of making liberal education feel like a subversive activity. This is because indoctrination, not education, remains the primary goal of Army, Navy, and Air Force alike.

Here, however, is where the Navy has the competitive edge. It alone has staffed and structured its academy to keep the mental and moral horizons of its graduates relatively unclouded by the uncertainties and ambiguities of liberal education. Naval Academy freshmen are called "plebes," and from the very first day of Plebe Summer, midshipmen learn that everything that will ultimately be important to them resides in "The Hall," not in "the Yard." "The Hall," in this case, is "Bancroft Hall," or "Mother Bancroft," the massive dormitory which houses the entire brigade and forms the locus of their military command structure. All places not "The Hall," in terms of a midshipman's mental landscape, are "the Yard"—an essentially alien place most commonly associated with academics. As part of their indoctrination, the mids are required to treat "The Hall" as if it were a ship. They must wear hats in the hallways and observe other naval customs associated with a ship underway. From this landlocked vessel, the mids are daily "put ashore" and sent out into "the Yard" on academic and other missions designed to test their mettle. But they remain accountable first and foremost to their upperclassmen and their company officers back in "The Hall." A mid's class standing and eventual opportunities, in fact, depend primarily on pleasing this military hierarchy. The academic virtues of free inquiry and respect for divergent opinion, for the most part, are daily put on and taken off like the uniforms they change according to the season and the occasion. The ethic that the mids imbibe back in "Mother Bancroft" includes the traditional military virtues of unwavering loyalty, unquestioning obedience, and decisive action. The message they hear loud and clear is that the Navy wants doers, not thinkers.

One of my civilian colleagues, a young assistant professor who routinely challenged midshipmen to examine their values and beliefs, for instance, once got this comment on his student evaluations: "*Professor X* seems like a good teacher, but his reading list is not consistent with Naval Academy values." *Professor X* lasted only two years at the academy. In my own case, I was called on the carpet in my first semester for encouraging my students to question the academy's commitment to equal opportunity for women. A female mid reported me. Even more disheartening, my fellow officers were not sympathetic. One even reminded me that, as a military professional, I should have realized that academic freedom did not extend to command policies and priorities.

A School for Scandal?

Mids also receive mixed messages when it comes to personal integrity. West Point holds its cadets to a clear and uncompromising "Honor Code": "A cadet will not lie, cheat, or steal, or tolerate those who do." The Naval Academy, on the other hand, has an "Honor Concept," admitting shades of gray and even giving mids the option of counseling one another instead of reporting an observed or suspected infraction. This is the point at which contemporary academic culture and Naval Academy culture almost meet, a single instance of plagiarism or misconduct generally being insufficient cause for expulsion at a civilian university. But the fact remains that the cadets at West Point and the mids at Annapolis are being groomed for positions of great responsibility. They should be held to higher standards than their civilian counterparts. It is not that the Naval Academy subscribes to the view that "to understand all is to forgive all." But its relatively flexible approach to honor has no doubt contributed to the widespread academic cheating and individual misconduct scandals that have made national news in recent years, setting Navy apart from the other academies.

In 1997, following a series of major, nationally publicized embarrassments, the academy formed a special commission headed by USNA alumnus and former CIA director Admiral Stansfield Turner. This commission, in its report entitled "The Higher Standard," largely shrugged off the suggestion that the academy had become a "School for Scandal." In the commission's view, there was nothing systemically wrong. All the problems leading up the commission's formation—ranging from sexual harassment and cheating to car theft rings and even murder—simply pointed up the need for more careful screening of applicants (Report 1997).

In keeping with the title of their report, the commission reminded the country that service academies exist to wean a select few away from the mores and values of mainstream America and to socialize them in the "higher standard" of unquestioning loyalty and selfless service. The commission took as a given that the Naval Academy is indeed morally superior to the permissive society which it serves and that, inevitably, a few mids will not live up to the "higher standard" of conduct and integrity the academy demands. What the commission left unsaid, however, is that the mids who rise to this "higher standard" cannot help but look down on their seemingly weak-willed, self-indulgent counterparts in civilian universities—an attitude the military hierarchy, in my experience, does little to counter and much to encourage.

"Too early a moral judgment," Harvard epistemologist William G. Perry, Jr. (1963), has written, "is precisely what stands between many able students and a liberal education." The subculture of "The Hall" still pays homage to the image of the ideal military leader as a taciturn man (or woman) of action intolerant of criticism or dissent. During my time at the academy, I heard mids as well as officers quote again and again from Theodore Roosevelt's "Man in the Arena" speech:

> It is not the critic who counts; not the man who points out how the strong man stumbles, or where the doer of deeds could have done them better. The credit belongs to the man who is actually in the arena, whose face is marred by dust and sweat and blood; who strives valiantly; who errs, who comes short again and again, because there is no effort without error and shortcoming; but who does actually strive to do the deeds; who knows great enthusiasms, the great devotions; who spends himself in a worthy cause; who at the best knows in the end the triumph of high achievement, and who at the worst, if he fails, at least fails while daring greatly, so that his place shall never be with those cold and timid souls who neither know victory nor defeat. (Roosevelt 1910)

Having committed themselves to lives of service and action, the mids are encouraged to think of themselves as soon to enter the "arena" and therefore superior to indecisive sorts who think too much and act too little. Fortunately, thanks to the efforts of the English faculty, some do learn the wisdom of living life on the balls of your feet from Hamlet instead of John Wayne.

This is not to say that the Naval Academy succeeds in turning all, or even most, of its midshipmen into arrogant, self-righteous, anti-intellectual prigs. But the fact remains that all the academies espouse another article of faith that contributes to the smug sense of superiority they encourage among their "long grey" or blue lines. This is the fallacious *post hoc* conviction that only their programs could have produced the great World War II generals and admirals who defeated the original Axis of Evil and preserved our way of life. The proponents of this belief ignore the fact that, before 1940, there were very few avenues to commissioned rank outside the academies. Who is to say that non-academy counterparts to Chester Nimitz, William "Bull" Halsey, Dwight Eisenhower, or George Patton could not have risen to the occasion? The academies have also certainly turned out their share of officers with questionable abilities and judgment (George Custer and William Westmoreland come to mind). It should also be remembered that our top officer in World War II, General George C. Marshall, was not a West Point graduate. Rather, he graduated from an academy dedicated to the civilian-soldier ideal, the Virginia Military Institute. In more recent years, many of our military superstars rose from the non-academy ranks. General Colin Powell is a case in point, and "three of the last six chairmen of the Joint Chiefs of Staff" were ROTC graduates (Ricks 2009).

Diversity and Inclusion

Articles of faith aside, however, the most bitterly contested intersection of military and academic culture, in the service academies and beyond, remains the Department of Defense's "Don't Ask, Don't Tell" policy. Unfortunately, during my time at the academy, the most disheartening and troubling assault on the traditional academic commitment to free expression and free inquiry I have ever witnessed occurred over this policy. During an ostensibly free question-and-answer period, a midshipman had the temerity to ask a visiting Secretary of Defense how much longer the Department of Defense was going to maintain the "old chestnut" that gays could not serve effectively as members of the armed services. This mid was roundly booed and was later accosted by so many of his classmates that the academy seriously considered placing him in protective custody. Worse, because of his impertinence, he was summoned the next morning for a private dressing-down by the superintendent himself.

I doubt that today's midshipmen would react so vehemently to a challenge to "Don't Ask, Don't Tell." A 2006 Zogby poll found that the majority of today's troops would be comfortable serving alongside gays (Palm Center 2006), and no less than General John Shalikashvili, former Chairman of the Joint Chiefs of Staff, has gone on record as no longer opposing allowing gays to serve openly (2007). Still, the policy remains in force and has led to considerable friction between the military and liberal universities committed to diversity and inclusion. The resolution, I predict, is soon to come now that the Global War on Terror is being fought on two fronts and our All Volunteer Force is being stretched to the breaking point. The Department of Defense will have to bow to expediency over the gay issue no less than academia has had to make its accommodation with the marginal level of preparation, as well as the manners and mores, of today's students.[11]

The current war has likewise helped women come a long way toward gaining full acceptance and equal opportunity in the military. It wasn't until 1976 that women were finally admitted to the service academies, and as former Naval Academy professor Carol Burke (1992) has pointed out, they have had to endure grudging acceptance on the part of an administration forced to bow to political pressure as well as open antagonism and actual harassment on the part of their male classmates.[12]

---

11 Recent developments in 2010 seem to indicate that this is indeed the case.

12 I reported to the Naval Academy a year after a group of male midshipmen forcibly handcuffed a female midshipman to a urinal and then proceeded to photograph her in that predicament. Academy officials disingenuously portrayed this incident in terms of good-natured, normal adolescent hi-jinks and not as a statement of a fundamental moral inadequacy. The choice of a urinal, moreover, argues against this gender-neutral interpretation of the motives for the handcuffing (Johnson and Harper 2005: 221-222). Since the act was not premeditated, academy officials decided, it did not constitute hazing. The penalties were limited to restriction, demerits, and written warnings. The female midshipman resigned in frustration over the academy's handling of the investigation (*Toledo Blade* 1990).

Here, in particular, the ghost of James Webb can be heard to rattle its chains. Women in that first coeducational class, the class of 1980, suffered a major setback in 1979. That was the year in which Webb published an opinion piece in *Washingtonian* magazine entitled "Women Can't Fight." Webb argued that, "by attempting to sexually sterilize the Naval Academy environment in the name of equality, this country has sterilized the whole process of combat leadership training, and our military forces are doomed to suffer the consequences." Conservatives soon began to chime in with Webb, warning that the country would recoil in horror at the spectacle of women returning from war missing limbs or, even worse, in flag-draped aluminum transfer cases. The fact is that all this has happened, and there has been nary a note of protest from the country at large (Palm 2008). While women are still officially barred from the infantry and a few other branches of the combat arms, they are now flying helicopters, driving trucks, and serving in other direct combat support roles that place them in harm's way. Women have proven that they can fight, and Webb (2006) has since rattled chains of regret over the extent to which he may have contributed to making the transition harder than it had to be.[13]

Military Monasteries

Thomas Ricks, *The Washington Post*'s noted military reporter and senior fellow at the Center for a New American Security, has gone on record as calling for the closing of the three federal service academies. Ricks points out that we spend over twice as much in educating an academy officer as we do an ROTC graduate ($300,000 versus $130,000). He argues that our money would be better spent on expanding the ROTC scholarship programs. Ricks singles out West Point in particular for not giving us our money's worth. This is because most of the instructors hold only master's degrees and the cadets, as a consequence, are essentially getting "community-college educations." "Some commanders" he has talked to, Ricks adds, prefer ROTC graduates "because they are better educated and less cynical about the military" (Ricks 2009).

Naval Academy officials and graduates can easily answer Ricks's concern for academic credentials. The preponderance of Navy's academic faculty members are civilians who hold doctorates. Ricks's argument, however, is concerned about much more than the expense. He suggests that the military and our society as a whole would be better served by officers who are educated alongside students destined to become professionals and leaders in other walks of life.

In the final analysis, the strongest argument for closing them is neither economic nor academic. It is ideological. Over a decade ago, in the pages of the *Atlantic Monthly*, Ricks warned that our All-Volunteer Force was growing less

13 See also Webb 1987 for his Forrestal lecture, given at the Naval Academy.

representative of American society at large and more intolerant of our democratic traditions of freedom of expression and dissent (Ricks 1997).

Ricks's concern certainly resonates with me. I personally heard more than one officer express admiration for the decisive manner in which the Chinese government had dealt with dissent in 1989 at Tiananmen Square. Our civil-military estrangement was becoming especially pronounced in the 1990s, when all the services were faced with declining budgets. At that time, every branch of the service was cutting back on ROTC and OCS programs and opting to place a heavier reliance on the service academies for officer accession. This trend, of course, has lately been overtaken by events. Two wars have forced the Department of Defense to call up large numbers of reserve and National Guard troops, making our "All-Volunteer" force more representative of our society than it otherwise would be. But still, as Ricks certainly realizes, the academies exist to indoctrinate more than educate and that pursuit runs counter to the ideal military-civil relationship in America.

While it is never openly stated in their promotional literature, the federal service academies are military monasteries. They cloister and mold their cadets and midshipmen apart from the influences of our permissive and self-indulgent society. This is why personal liberty is restricted at all three of the academies, especially during the first year and to a lesser extent in the second and third years. A fellow officer, himself an academy graduate, once summed up the *raison d'être* and the strategy in the following terms: "We need to capture them while they're still young and impressionable." The academies seek to develop a strong emotional attachment and a sense of personal identification with the military. In this sense, they really do exist to foster military spiritualism. The academies hope to turn out officers who not only view the military as a profession but who embrace it as an all-encompassing way of life—in much the same way that some people form strong emotional bonds with the religions in which they were raised. The problem with this is that true believers—in the religious or the military sense—have a tendency to take a pharisaical pride in their commitment and to close their minds to other points of view.

The American military does not need ideologues. It needs officers with a spirit of intellectual independence and whose frame of reference transcends the institutions they serve. And the surest method of achieving this, in my experience, is through ROTC.

Winning Hearts and Minds at Berkeley

There is no question that ROTC has been, and remains, a contested site on liberal university campuses. Academics generally do not embrace ROTC units as an opportunity to form an alliance between Athens and Sparta. To the contrary, they typically view ROTC as an encroachment. ROTC staff members, in turn, tend to reciprocate by keeping a low profile and holding themselves aloof from the campus at large. Given that the great majority of academics today have had no military experience, a certain degree of friction is understandable, if not inevitable. An ROTC unit, after all, represents a kind of colonial project—an attempt to convert a certain number of the "natives" to the manners and mores of a different culture. What both sides sometimes fail to appreciate, however, is that neither culture—academic or military—is "better" than the other. They are just different, and there are, in fact, more intersections between the two cultures than are commonly recognized. As colonial projects go, therefore, ROTC is one that the military should fully support and which the academy should welcome as an opportunity for constructive engagement and mutual understanding.

Before being assigned to teach English at the Naval Academy, I served for three years—from 1984 to 1987—as the Marine Officer Instructor assigned to the Naval ROTC unit at the University of California, Berkeley. I taught military history and military affairs and oversaw military training, but from the Marine Corps's standpoint, my major responsibility was to serve as a role model and to interest qualified midshipmen in opting for commissions in the Marine Corps rather than the Navy. Knowing my academic credentials and interests, people have been surprised to learn that I look back at my time in Annapolis as the three worst years I spent as a Marine officer and my three at Berkeley as the best.

My reasons for feeling this way are partly personal but mainly professional. My own résumé includes service as an enlisted Marine in Vietnam, but in a very real sense, everyone who came of age during the Sixties is a Vietnam veteran. The shooting war "in country" was only the manifestation of the larger cultural contest being waged back here—back in "the World," as those of us who served "in country" used to say. The decisive theater turned out to be back here in "The World." This outcome remains a source of residual bitterness for some of my fellow veterans, but not for me. I went on to study the history of our Vietnam involvement and to become a student of the ethical implications of the American cultural narrative. Early on, I came to terms with the fact that all of us who went to Vietnam had borne the brunt of friendly folly as well as enemy fire. I further came to realize that, while the motives of those who protested may not have always been selfless, the antiwar movement and the press had ultimately done us a great service in helping to force an end to the original "wrong war in the wrong place at the wrong time." I therefore embraced my Berkeley assignment as an opportunity to step back in time and to meet some of the people who had manned the other side of the ideological barricades.

My entrée into that world came about through an ironic conjunction of chance and circumstance. Callaghan Hall, the home of all three ROTC units on the Berkeley campus, had withstood 20 years of anti-ROTC sentiment and protests, only to fall victim to an arsonist in the early morning hours of February 18, 1985. The building was a total loss. An invaluable collection of personal papers and rare books that had belonged to legendary World War II Admiral Chester Nimitz (who had started the unit at Berkeley) was also lost—along with irreplaceable personal papers and memorabilia belonging to all of us then assigned to Berkeley's ROTC units. For me personally, however, the most disturbing aspect of the whole experience occurred a few days after the fire, when we were waiting for the Fire Marshal to allow us in to the burned out shell of the building to salvage whatever we could of our personal effects. A very young, angelic-looking woman who happened to be passing by said "I'm glad it burned!"

That remark stuck in my craw—so much so that I penned a somewhat bitter op-ed piece for the *San Francisco Chronicle* aptly entitled "Book Burning at Berkeley." The main theme of that piece was how little the campus at large knew about who we were and what we stood for. In addition to recounting who I was and how I came to be there, I made the standard case for ROTC as an important tempering influence on our military establishment. I also pointed out that anyone who could delight in our loss had dehumanized and relegated us to roles in essentially the same psychic melodrama that Berkeley's Peace and Conflict Studies Department held to be largely responsible for war in the first place (Palm 1985a).

As a result of that article, the editor of *The Daily Californian* invited me to take part in a point-counterpoint exchange on the merits and academic legitimacy of ROTC. In my piece, I countered all the hysteria, hyperbole, and demagoguery then held to be valid arguments for kicking ROTC off campus (Palm 1985b). I assured the readership that we were not secretly recruiting officers for intervention in Central America and that, like the American military in general, we were politically neutral. I reminded everyone that the principle of civilian control of the military is sacrosanct in this country, but that enlightened military leaders have on occasion advised against military adventures they considered to be reckless and not in our country's best interests. I acknowledged that members of the military do forfeit some of the rights enjoyed by civilians, but only those necessary to ensure the discipline and effective functioning of military units. In answers to a specific charge then being leveled against us, I reassured all concerned that ROTC cadets and midshipmen were free to organize and express their opinions on any issue or topic, so long as they made clear they were acting as individuals and not speaking for the Department of Defense. I addressed the issue of personal coercion, stressing that no one is forced to join ROTC and that scholarship and non-scholarship students alike are given a reasonable trial-period before incurring any obligation. I tried to put the most emotional argument in some perspective, explaining that no one is "trained to kill" within the context of the ROTC program and that the great majority of servicemen and women are assigned to non-combat roles. But the most important issue I addressed, to my mind, was whether ROTC constitutes a legitimate academic program.

My answer: it can and should be, and on the Berkeley campus in the mid-1980s it indeed was an academic program. All three ROTC programs—Army, Navy, and Air Force—were organized into one department, and all three had agreed to require an interdisciplinary, team-taught course unique to Berkeley's ROTC curricula. Called "The Military in American Society," the course covered topics such as the history and organization of our military establishment, the concept and character of the military profession, the nature of the civil-military relation in America, the causes of war, the Just War Theory, the law of land warfare, and the issue of personal responsibility versus military discipline. The readings for this course included articles and excerpts from books by leading military historians and sociologists and by journalists and others critical of the military. We also offered a full array of courses in military history, some of which were required by all three services and all of which were taught from an objective, apolitical standpoint. Our academic credibility rested on the fact that we exposed midshipmen and cadets to a variety of viewpoints and opinions, and we challenged them to question and to think for themselves—especially about whether the military was right for them. Our courses, moreover, were not insular enclaves of like-minded individuals. They were open to the campus at large, and non-ROTC students did routinely enroll, resulting in frequent lively exchanges.

Unfortunately, not all my fellow officers were supportive of requiring academic courses over and above the professional military courses prescribed by their respective services. One senior officer was even heard to say that Berkeley was, in effect, "rubbing our noses in My Lai." This comment, coupled with the trauma of the arson fire, prompted me to write "Winning Hearts and Minds at Berkeley," an article I published in the Marine Corps's professional journal which won me a certain notoriety among my fellow Marines.

In that article, I argued for wearing uniforms only on drill days or for other purely military functions. In donning uniforms, I pointed out, we step into an institutional persona. We stereotype ourselves and inhibit the sort of open interaction and free exchange valued in academe (how many people, for instance, are comfortable talking to a uniformed police officer?). I argued for de-emphasizing close-order drill, as officers typically do not engage in it, and the spectacle of uniformed students marching in ordered rows gives the erroneous impression that ROTC indeed seeks to turn students into mindless automatons. I recommended that military training in general take place off campus, during the summers and after commissioning. But, above all, I argued that we needed to get serious about academics. I conceded that ROTC officers had to be acceptable as military role models, but I also insisted that those role models be well-versed in military affairs and military history and that they possess the academic credentials to teach these disciplines on the college level (Palm 1986a).

My teaching, my articles, and my general willingness to interact with the university community soon won me a reputation as "the liberal Marine." On more than one occasion, I found myself pushed out front, figuratively as well as literally, to defend and explain ROTC to its detractors.[14] It was in this role, as a

14 The high point of my career as an apologist for ROTC came in the spring of 1987, toward the end of my tour of duty at Berkeley. The university's alumni magazine, in its June issue, featured me and Professor John Hurst, head of the Peace and Conflict Studies program, in an article entitled "Polar Bears." The author of the article, Timothy Pfaff, pointed up the parallels between ROTC and PACS and allowed me and Hurst both

self-appointed defender of our faith in ROTC, that in the spring of 1987 I found myself detailed to meet with legendary Sixties musician and activist Country Joe McDonald of Woodstock fame. Country Joe was then trying to organize a week-long Vietnam Film Festival and Arts Fair, which he and the Associated Students of the University of California were billing as "a tribute to those who fought and resisted the Vietnam War." Having contributed to the polarization of the country in the Sixties, Country Joe, in my estimation, sincerely wanted to erase the bitterness of the past and to bring us back together, both those who protested and those who fought. And he very much wanted the military on campus to join in the festivities.

In the end, the leaders of all three units were chary of taking part in these events. But I attended privately, and I went on to take part in a number of panel discussions on Vietnam with Country Joe and other activists. I found that I felt a certain kinship with those who had protested and resisted. After all, we had the Vietnam experience in common. This was more than I could say for most of my fellow officers. Some were too young to have served or protested. But, worse, others had intuited and internalized the attitude of a military establishment still in denial about our defeat and about its strategic, political, or cultural implications.

At the time, an astute Army officer of my acquaintance put it best: "It's OK to be a Vietnam veteran in today's military, so long as you don't dwell on it or refer back to it." The military establishment of the 1980s was eager to dismiss Vietnam as an aberration and to resume preparing and training for conventional, linear warfare. I saw it differently. I felt that it was our responsibility as military professionals answerable to civilian authority to confront not just our tactical failings but also the political, cultural, and moral reasons why we had not prevailed in Vietnam. What better place to do that than on the campus of a liberal university?

I would hope that today's ROTC students, at Berkeley and elsewhere, are likewise being challenged to think long and hard about the ethics and responsibilities of the military profession as well as the whys and wherefores of our current wars. What I gained from teaching at both Berkeley and Annapolis, finally, was a new-found appreciation for Milton's refusal to "praise a fugitive and cloistered virtue."

The Naval ROTC mids at Berkeley were not only the most intelligent, interesting, and diverse group of students I have had the privileging of teaching; they were also open-minded, tolerant, and adept at moving between the academic and military cultures. Like the Naval Academy, we too encouraged our mids to identify with our unit, but as their fraternity rather than an all-embracing ideology or way of life. The crucial difference is that our mids were exposed daily to all the temptations that flesh is heir to and which a major university can provide.

to make strong cases for the academic legitimacy of our programs. Rather than being commended for my stalwart defense of our program, I was counseled that it would have been more appropriate for the head of our unit, the Professor of Naval Science, to emerge as the principal spokesman for our program. My understanding is that he was contacted but declined to contribute to the article.

They were not nurtured in an insular culture and surrounded only by like-minded individuals. Fellow students and professors confronted them with divergent opinion, and even outright hostility, about the role of the military in American society. Our mids were challenged to examine and defend their values and beliefs. As a result, we lost some students along the way. But the ones who stayed with us had made reasoned, well-informed commitments, and I have to believe they became better officers for the experience.

Having served in both programs, I believe that Ricks is right. The military is better served by winning hearts and minds at universities like Berkeley. There, at least, the cadets and midshipmen do not have to be conflicted, torn between liberal education and military socialization and indoctrination. They can pursue liberal educations, to borrow a phrase from Joseph Conrad, with a "singleness of intention," leaving their professional military education and training for later—which, of course, is the process favored by other highly responsible and complex professions.

This is not to denigrate Naval Academy alumnus and now-Senator James Webb, whose recent election campaign and longstanding opposition to the war in Iraq speak for themselves (as well as his willingness to rethink and recant his former opposition to women in the military). Webb clearly did not allow his Naval Academy schooling to interfere with his education. But out of his same class, the class of 1968, came Oliver North, whose past role in the Iran-Contra scandal and continuing association with right-wing causes and pundits would seem to validate all of Ricks's concerns, as well as mine.

The advent of permanent military professors notwithstanding, I hope that Webb's ghost continues to haunt the halls of the Naval Academy. It is "an honest ghost" with much to tell us about what's really rotten in Annapolis and why military officers need to be liberally educated.

Chapter 2
Combat Ethnography

Carol Burke

The US Army's Human Terrain System (HTS) fielded its first teams in Afghanistan and Iraq in 2006. The program is designed to enlist the help of academics with cultural expertise in fighting our current wars of counterinsurgency. These wars have invariably involved engagements with an enemy that looks just like civilians and civilians who, because of their fear of US power and their bewilderment at US intentions, often behaved like the enemy. Prior to deployment, these five-to-nine person teams of civilians receive four months of training at Fort Leavenworth, Kansas in which they learn how to acquire knowledge of the local culture, whether civic, tribal, or sectarian, and provide commanders with the guidance that would avoid tactical errors with strategic consequences: not whom to hit and where to hide but what matters and how to help. Or, as one commander put it to the human terrain team who was running through a training scenario at Leavenworth, one in which they had supposedly just met with a group of village elders, "I need to know if I give them a handshake or a hand grenade." Every team of civilian contractors includes one or two social scientists, a team leader with senior military experience, a research manager who helps design research projects and presents their findings, and analysts who may speak a local language, conduct interviews, or survey research materials.

Anthropology and the Military in the Twentieth Century

Military efforts to recruit academics with cultural knowledge are nothing new, nor is the blowback from academics who fear the effects of such collaboration on their discipline. In 1917, anthropologist Franz Boas denounced fellow anthropologists whose fieldwork in distant places served as a convenient cover for their spying activity.[1] In 1919, Boas wrote a letter to *The Nation* condemning the deception of these anthropologists and raising alarm about the lasting harm of such practices on their discipline. Anthropologists count on the trust of local populations, according to Boas, and when a group of anthropologists pretends to be collecting one form of information for one reason but is, in secret, collecting human intelligence in the

1 The anthropologist who has written most extensively about the collaboration of government and anthropological inquiry is Price 2008. His book begins with a discussion of Franz Boas' 1919 censure by the American Anthropological Association.

service of military commanders, they endanger the fieldwork of future generations of anthropologists. As a result of his strong denunciation, Boas was censured by the very professional organization he had helped bring into being, the American Anthropological Association. Boas was only the most famous anthropologist to suffer the consequences of stating his unpopular criticism; others faired worse and were actually fired from their university posts for their candor.

By World War II, calls for anthropologists to refrain from applying their skills to military purpose were muffled by the urgent cries from Washington for more and more cultural knowledge needed to win the wars in Europe and the Pacific. With the tacit approval of their professional organizations in World War II, American anthropologists rallied to a popular cause and flocked to the War Department to lend their expertise. They produced language training materials and phrase books for G.I.'s. They contributed to country "pocket guides," 4" x 5" booklets designed to fit neatly in the pockets of G.I.'s heading to exotic locales. These pocket guides typically included a brief history of the country and generalizations about the customs, manners, and religion of the people. They sought to caution the American soldier from behaving in ways that might be socially acceptable in the US but would be regarded as offensive by a foreign community. The World War II country guide to Iraq, for example, encouraged soldiers to shake an extended hand rather than slap the back of an Iraqi, to avoid eye contact with women, to refrain from drinking in sight of Iraqis, and to reserve cussing for fellow Americans rather than the Iraqis they might encounter: "Do your swearing in English … Don't under any circumstances call an Iraqi a 'dog,' a 'devil,' a 'native,' or a 'heathen.' These terms are deadly insults to him."

Anthropologists helped craft propaganda appropriate to the remote populations they had studied, and they assisted intelligence gatherers. They signed on to work with the War Relocation Authority, created in 1942 to intern 120,000 Japanese Americans in ten camps in Arkansas, Arizona, Colorado, Wyoming, California, Idaho and Utah. In order to increase the utility of their method, they taught officers the basics of anthropology. To aid in this effort, the War Department, no doubt ignorant of the irony, reissued Franz Boas' *General Anthropology*, as "Education Manual 226".[2] In 1943, Margaret Mead concluded that the War Department had already enlisted the help of most American anthropologists and that a longer war would produce a shortage of trained field workers, so she sketched out a plan that sounded a little like today's HTS. It would have establish a cadre of soldiers trained as "regional ethnographic specialists" who would be schooled in anthropology, geography, writing, local languages, history, and customs, and who would have served commanders in the field (Price 2008: 79). Few anthropologists questioned this massive shift from "pure" anthropology to "applied."

Application often entailed the passage from obtaining knowledge about a culture to intervening in the practices and altering the perceptions of indigenous peoples. While stationed in Burma, Gregory Bateson, Mead's husband and an enthusiast

2 Boas wrote the introduction and six of the text's fifteen chapters.

for the use of anthropological methods in assisting the Office of Strategic Services (OSS), learned that for the Burmese the color yellow marked the end of a period of foreign occupation, so he set about dropping dye into the Irrawaddy River, the country's most important commercial waterway. If he could not part the waters, he would at least change their color. He reasoned that the locals would interpret such a marvelous transformation as an omen and rise up and expel the Japanese. To his disappointment, Bateson discovered that although the dye he used would turn salt water yellow, it just sank to the bottom of the river with no perceptible change: no tint, no omen, and no overthrow.[3]

Although an important anthropologist, Bateson was a bit player in the use of anthropological methods in waging war. Major General Edward Lansdale, a former OSS member like Bateson and a former ad man, concocted several schemes in his counter intelligence career that were every bit as outlandish as Bateson's. While serving as intelligence advisor to the Philippine armed forces in the 1950s, Lansdale learned that the locals and the communist-inspired insurgents, the People's Liberation Army, feared that vampires ("Asuangs") might linger near a battleground. Just as Duvalier used Haitians' fear of zombies to control the superstitious poor, Lansdale hatched a plan to send insurgents on the run. He wrote about this and other incidents, which he refers to collectively as "the practical-joke aspect of psywar," in his memoir, *In the Midst of Wars: An American's Mission to Southeast Asia:*

> A combat psywar squad was brought in. It planted stories among town residents of an Asuang living on the hill where the Huks [the People's Liberation Army] were based. Two nights later, after giving the stories time to circulate among Huk sympathizers in the town and make their way up to the hill camp, the psywar squad set up an ambush along the trail used by the Huks. When a Huk patrol came along the trail, the ambushers silently snatched the last man of the patrol, their move unseen in the dark night. They punctured his neck with two holes, vampire-fashion, held the body up by the heels, drained it of blood, and put the corpse back on the trail. When the Huks returned to look for the missing man and found their bloodless comrade, every member of the patrol believed that the Asuang had got him and that one of them would be next if they remained on that hill. (1991: 72)

Unlike Bateson's attempt to turn the river yellow, Lansdale's efforts to turn ethnographic data into bizarre tactics actually worked. The following day the insurgents hightailed it out of the area.

From the Philippines, Lansdale moved on to counterintelligence in 1950s Vietnam, where he sought to learn as much as he could about the Vietnamese. As he had with the Japanese in World War II, Lansdale began collecting Vietnamese proverbs because he believed that they offered insight into the thinking of the people

3 Julia Child had, like Bateson, served in the OSS, and this anecdote first appears in a biography of Child (Fitch 1999: 100), and is later mentioned in Price (2008: 239).

that would be useful in winning the war. He collected Vietnamese folksongs, jokes, and Vietnamese political slang. He even published his own astrological almanac in 1955 and arranged for it to be smuggled into the hands of the enemy so that the soothsayers who advised commanders would use his almanac. To his surprise, the almanac went into a second printing, making an impressive profit.[4] Lansdale's insistence that an occupying force could not defeat its opponents without first understanding them and their culture led eventually to a war for the hearts and minds of the people, which operated along side the "kinetic" war in Vietnam. The Civil Operations and Revolutionary Development Support Program (CORDS) began in 1966. A program on a much larger scale than today's HTS, CORDS brought together civil and military assets and deployed them to rural areas of the country, in operations like USAID, CIA, Psychological Warfare, Civil Affairs, and the Military Assistance Command, Vietnam (MACV).

Jim Bullington, a career diplomat who was a young aide to Ambassador Henry Cabot Lodge when he joined CORDS, recalls a robust program with team sizes of 150 military and civilian personnel, as opposed to today's 5-9 person Human Terrain teams. Before CORDS, said Bullington, "the previous attitude had been if you grab him by the balls, the heart and mind will follow. In CORDS, we geared everything towards developing in the populace a reason to support the government."[5] Bullington faults the US military for taking 40 years to admit that there were both strategic and tactical lessons to be learned from the Vietnam War, particularly from what he calls "the other war," the war, he insists, that America actually won in Vietnam, the war of counterinsurgency. His advice to those waging today's wars of counterinsurgency: "Listen to the prophets like Lansdale." Bullington believes that today's HTS vindicates the approach Lansdale creatively employed first in the Philippines and then in Vietnam. "CORDS was respectful of the culture," he insists. "You didn't disrupt things. You concentrated on developing in the populace reasons to support the government."

The Human Terrain System in Theory and Practice

CORDS and the HTS have both focused on winning the support of the local population, but that's where the similarity ends. In CORDS the teams lived among the locals. Members of the HTS teams live on base and go "outside the wire" for short interactions with the local population. While most of the CORDS personnel were operating like Peace Corps workers in the jungle villages of Vietnam, their lethal counterpart, the infamous Phoenix Program, sent hit squads out to assassinate suspected insurgents.

4 For more on Lansdale's ethnographic interests and especially on his large collection of occupational folksongs, see L. Fish 1989.

5 Bullington's quotes and impressions of CORDS were obtained through a phone interview with the author on July 11, 2008.

The idea for the army's HTS came about well after the invasions of Afghanistan and Iraq. Anthropologist Montgomery McFate was working for the Navy and was asked to create a database that would include both historical material and contemporary information relating to the culture of the populations in current areas of conflict. When done, she met with a group of colonels at the Pentagon, of whom Steve Fondacaro was the most outspoken. What he didn't need, he told her, was another database: "We have oceans of data, but when a commander is trying to answer questions, he doesn't need the Pacific Ocean, he needs a glass of water … Bits and bytes don't yield information; it takes human expertise to come up with that. When you further refine the information, it becomes knowledge, and then you can achieve understanding. Then I have what I need."[6]

Although HTS's organizers had hoped to recruit anthropologists willing to lend their skills to the fight, academic anthropologists, eager to repudiate their discipline's colonialist past, became the program's first vocal critics. Many voiced the charge that the program threatened to breach the ethical code of anthropologists who ask, how can informed consent be maintained in a war zone where the "ethnographer" is backed up by a soldier with a rifle? How can one promise to "do no harm" when, in some circumstances, even conversations with Americans in uniform may be viewed as collaboration and reported to those eager to mete out rough justice? HTS managers have replied to these claims by arguing that human terrain teams provide hard facts about who the locals are, how they live, what they hope for, and what they fear. They claim that by encouraging military units to be more cautious in their decisions to deploy lethal force, these teams actually save lives.

The team members have not been shy about voicing their own criticisms of the program. They have complained about erratic leadership at both the local level (team leaders who don't lead) and back at the program's headquarters at Fort Leavenworth. They fault managers so eager to fill positions that they have recruited individuals with little experience negotiating in hostile areas or unfamiliar with fieldwork and the swift but judicious decisions about what to collect of village life. As of April 2009, only 49 of the 417 employees of the program held PhDs in anthropology (American Anthropological Association 2009: 12). The average time-to-degree for PhD recipients in anthropology is ten years, according to the National Science Foundation. A master's degree typically requires two years. And some of the program's social scientists come with no prior experience in fieldwork.

Some team members have criticized the four-month training program for failing to adequately prepare them for the tasks they faced "downrange." The Army is currently assessing its training program in response to this criticism. Other team members have complained of ill-defined job descriptions and little coordination among teams. HTS teams are not the first groups of civilians working with the Army and the Marine Corps to experience problems of cohesion and

6 Comments were made in a phone interview with the author on October 14, 2008.

coordination. The State Department's Provisional Reconstruction Teams (PRTs) have encountered similar difficulties. In a December 2008 report evaluating Iraq's PRTs, Rusty Barber and Sam Parker of the US Institute of Peace found that if the leader of a PRT is strong, the mission well articulated, and the PRT filled with professionals, the team jells and brings to the field invaluable expertise. If, on the other hand, teams lack strong leadership, cohesion, and experienced professionals, they often break into a group of freelancers who opt to pursue projects that are neither well thought-out nor coordinated with the work of other teams in the region or the country (2008).

Engaging with the local population in the middle, or even the tail end, of a war is not easy. The members of every HTS team who hear the nightly mortar attacks know that they are risking their safety even when hunkered down on the base. The members of every team know too that the risk they take increases a hundred fold once they venture outside the base. They know that when their formidable convoy of MRAPS, those $800,000 vehicles that soldiers call "Humvees on steroids," enters a village and they dismount from their mobile fortresses, they must heighten their "situational awareness," get the subjects, get the facts, and get out. Nothing makes the military units who take them off base more nervous than sweating out long moments in a street that might have trained and hostile militants just around the corner.

Soliciting cultural information from a potentially hostile population to be downloaded onto a computer back on base is not a task that the US Army has trained for, except for those who gather "human intelligence." Human intelligence is information obtained for an objective that makes clear military sense: determining targets. The rules of the game have changed in Iraq and Afghanistan. Because of the shift from a mission of "search and destroy" to one of "clear, hold, and build," commanders must realize that "the key to success in counterinsurgency is protecting the population," according to John Nagl in his forward to the army's *Counterinsurgency Field Manual* (United States Army and Marine Corps 2007: xv).

HTS was devised to acquire the kind of fine-grained knowledge of communities that escape aerial surveillance and infantry sweeps. A long war of counterinsurgency with no frontline and no way to quantify "kill ratios" is *lived* both by the occupying force and by the enemy. The latter is typically not an organized force but a network of cells and bands, distributed throughout a neighborhood or a region. According to McFate, the program's senior anthropologist, "Successful counterinsurgency depends on attaining a holistic, total understanding of local culture. This cultural understanding must be thorough and deep if it is to have any practical benefit at all" (2005: 38). Until the program can recruit seasoned ethnographers who have conducted research over long periods of time in the area of the fight, the "thorough and deep" understanding of a local culture will be elusive. The "holistic" understanding of local culture that we typically call ethnography can only be produced after sustained immersion, which is simply not feasible during this or probably any war. People tend to be in too much of a fright or a rage to sit down for long talks. As a result, HTS teams conduct "windshield ethnography," quick observation while passing through, short surveys, and "REAP"s (Rapid Ethnographic Assessment Procedures), with an emphasis on the "R." One HTS social scientist, who wished to be identified simply as "Nancy,"[7] describes her conclusions after a 4-hour REAP in which she and one other team member collected impressions from villagers in a town in the north of Iraq:

> There were absolutely no opportunities for employment in the town. Even subsistence farming was failing due to drought, lack of seeds, fuel, and water that was too salty and was killing plants. The water situation was terrible—a lot of sick kids. The town had the most garbage in it than I had ever seen. These were people with not a lot of hope.[8]

Typically, team members ask the locals whether essential services like water are adequate, whether they have access to health care, whether schools are equipped and students attending, and whether the local economy is functioning. When NGOs are operating in the area, they ask the people whether it is safe to engage with the NGOs and whether they are meeting their needs.

The HTS program attempts to make up for the lack of time and lived continuity by providing "reach back" capability. The team's social scientists can request reports on specific topics from headquarters, otherwise known as "Building 48" back at Fort Leavenworth, and access background information gathered by the Department of Defense, the State Department and the CIA and collected on classified sites, including patrol reports. Still, it's a patchwork. Teams need to piece their views of a culture together from short research projects that may not yield McFate's "total understanding" of a culture but that, nevertheless, may offer useful answers to reasonable questions about one significant aspect of the local culture.

To access the reports available through their "reach back" service, team members must have security clearance. Leaders of the HTS have insisted that the program has nothing to do with intelligence gathering. A year ago, however, the director of the program, Colonel Fondacaro, proposed renaming their positions as "intelligence analysts." In a hostile environment in which intelligence protects American forces and targets the enemy, it would be foolish to assume that information obtained by an HTS team would not be shared with the intel people. And that's how it worked when I was there. Before heading out into the field, the HTS team with whom I was embedded ran their questions by an intelligence officer at brigade headquarters. And while his fellow team members were out in the field, one team member stayed behind to assist the intelligence people on a project.

7 For part of December 2008 and January 2009, the author was an embedded journalist with an HTS team in northern Iraq. Nancy was the team's only social scientist.

8 [Email] Message to Carol Burke, January 31, 2009.

The teams are eager to please. After all, the value of HTS teams is not decided by a review committee at the National Institute of Health or the National Institute of Science; it is determined by the commanders whom they serve, those who assess their value in relation to the military objectives set for the mission. Those standards change quickly because no one stays long. Every time a new command cycles through (for the group I observed, it was three times in six months) the team must pitch itself to the incoming brigade staff, selling them on the expertise they bring to war.

Toward a More Effective Combat Ethnography

Officers in today's army, who have been trained to fight high intensity battles, are now called on to act as diplomats. They must be what General Petraeus has called "pentathlete officers." Some have mastered the skills and manage to swing the deals they need to get local tribal support. Others prefer to stay in their tanks. But why not flush them out? All officers report to commanders, who have come up through their commissioned ranks from West Point, officer candidate schools, and ROTC programs that have trained them in military duties and skills. Current wars have proven that not just cultural sensitivity but also cultural expertise among the officers in the field is vital. Yet there is no formal training in anthropology available for the elite officer class. West Point has a good English Department and a terrific debating team. Isn't it time that Sparta on the Hudson developed a department of anthropology?

Some members of the military would like to see more of a shift than simply moving these HTS teams into auxiliary positions. Because of the importance of the mission, they argue, HTS teams should be integrated into the army rather than "attached" to it. The Army would not need to make much of an adjustment to absorb the functions of the civilian teams. Military Intelligence, Psychological Operations, and Civil Affairs all focus on local populations. Military intelligence sits firmly on the lethal side of military operations and concentrates on targeting. Whether they pay informants to report on what is preached at the local mosque or monitor suspicious activity from satellites, they seek to separate the bad guys from the rest of the population. Psychological Operations typically strives to understand enough about the local population to design effective propaganda, and their communication typically travels one-way. Civil Affairs, however, has a long history on the non-lethal side of military operations. It feeds and shelters civilians when the Area of Operations is "highly kinetic" and works on reconstruction in times of low intensity conflict. Civil Affairs personnel meet with local councils, with local business owners, and with local farmers to help reestablish the civil and economic life that war has ripped away. As a Civil Affairs officer who wished to remain anonymous said, "We're the herbivores; they're the meat eaters. The army breaks shit, and we go in and make everybody happy."[9] Filled with reservists,

9 Interview with the author.

Civil Affairs brings to the battle doctors, lawyers, teachers, business owners, electricians, and other skilled workers who have spent more of their adult lives making their livelihood in the civilian world than in uniform. The problem is that there are just too few of them. Faced with limited numbers (together Psychological Operations and Civil Affairs total only 10,000 soldiers, 96 percent of whom are reservists) and multiple deployments (I met one Civil Affairs officer who had been deployed seven times), they are already over-burdened.

Whether the problems of the fledgling HTS are fixed, or whether the Army and Marine Corps start training officers familiar with the culture of the lands they occupy, the need for such knowledge will continue. Only in the unlikely event that we cease fighting wars of counterinsurgency and return to wars in which the only terrain to master is geographic will we cease to need this intelligence. Without the regiment of military anthropologists imagined by Margaret Mead in World War II, the Pentagon will need the help of academia. In a speech before the Association of American Universities on April 14, 2008, Secretary of Defense Robert Gates explained the rationale for the Department of Defense's "Minerva Research Initiative Awards," named after the deity with the twin titles of goddess of war and goddess of learning. He explained that:

> In Iraq and Afghanistan, the heroic efforts and best intentions of our men and women in uniform have at times been undercut by a lack of knowledge of the culture and people they are dealing with everyday—societies organized by networks of kin and tribe, where ancient codes of shame and honor often mean a good deal more than "hearts and minds." (United States Department of Defense 2008)

He spoke about the new threat facing this country, one that requires not just the "guns and steel of the military" but also "'soft power'"—"from diplomacy to economic development and assistance, institution-building, strategic communications and more." The way to unlock that "soft power," according to Gates, is to "find untapped resources outside of government—resources like those our universities can offer." Although the academy may have resources untapped by the military, the Minerva Initiative's miniscule budget (less than $3 million) seems paltry when compared to the total Pentagon's $708 billion budget for the next fiscal year. At a time when university budgets are shrinking, hiring freezes are going into effect, and academics are taking pay cuts, funding for basic research in the social sciences and humanities, for the study of less commonly taught languages, and for scholarship on the causes and cures of insurgency are certainly attractive offers to make to academia. But for a real partnership to be formed, government must involve academics in the planning and implementation of funded projects, ensure that the ethical standards that define a discipline are not violated, and subject joint ventures to peer review. Since World War II, the Department of Defense has invested lavishly on research in the hard sciences. If "soft power" is what is needed to secure peace in the world's volatile spots, then more than token funding must be forthcoming.

Chapter 3
An American Professor with the Iraqi Army[1]

Andrew H. Myers

"I request an extension of two weeks for my review," said my email to the journal editor. "Yesterday, the Iraqis I advise got hit with mortars and suffered two wounded. Today, my team had a nasty encounter with an improvised explosive device. One of my sergeants is going to be really sore tomorrow, and his ears are going to ring for a while, but he'll be okay. We were really lucky. You would not believe the kinds of paperwork these kinds of incidents generate, so I am going to have to set the review aside. I hope the delay will not overly inconvenience you."

"No problem," the editor replied. "I think yours may be the most plausible case for an extension I've ever received."

This exchange was one of many intersections of academic and military culture that peppered my fourteen months of active duty with the United States Army. As a civilian, I teach American Studies and history as an associate professor at the University of South Carolina, Upstate. As a Reservist with the rank of major, I led an eleven-man military transition team. From July 2006 until June 2007, we provided advice and assistance to an Iraqi infantry battalion located next to the town of Al Ba'aj (also spelled Biaj) in western Nineveh Province.

Unlike many Reservists, I knew my mission and approximate mobilization date more than two years in advance. My university colleagues supported me wholeheartedly. The staff at the human resources office made certain that my family received all of the pay and benefits that I would be authorized during the leave of absence. The editor at the press publishing my first book bumped the manuscript to the front of the line for copy editing. My department chairman arranged for someone to teach classes in my stead. My file for tenure and promotion received approval even though many members of the faculty opposed the war.

Working in an academic setting helped me use the available time to good advantage. I developed a list of books and started reading as I would with any new research project. I received help from a political scientist in my department who specialized in the Middle East and who had lived in Iraq. The kind of information she could provide was not readily available through Army sources during the early

1 The views expressed in this chapter are the author's and not those of the United States Army, the Department of Defense, or the United States government. All of the information presented is unclassified.

years of the war, and what little the military distributed was sometimes inaccurate. For example, our cultural awareness brochures insisted that Arabs take offense at people who eat with their left hands.[2] My colleague argued otherwise. Experience proved her to be correct.

I also began listening to Arabic language lessons during my daily commute. The Army did not offer any formal training other than flashcards and hard-to-hear diskettes available through the Defense Language Institute. I used my own money to purchase a civilian system. Being able to exchange basic niceties in Arabic helped, but overall this effort bore limited fruit because a significant portion of the Iraqi battalion I would eventually advise consisted of Kurds, who often responded to my attempts at speaking Arabic by trying to teach me the equivalent in their language.

I conceptualized the advising job as one of teaching as much as soldiering. The reserve institutional training division to which I belonged, the 108th, consisted of drill sergeants and instructors. I had spent fifteen years in this kind of unit and had served as a basic training company commander, division staff officer, Reserve Officer Training Corps instructor, and basic training battalion executive officer. Most of the men originally assigned to my team were drill sergeants who practiced their jobs one weekend per month and who normally spent two weeks out of every year working with privates. Although we anticipated some fighting, we viewed ourselves primarily as pedagogues. Why else would our unit have been mobilized? Combat advising, we thought, belonged to the realm of the Special Forces.

Conditions in Western Nineveh Province, 2003-2006

The first reserve institutional training division to have this mission, the 98th, was deployed during late 2004. Its advisors found themselves not in classrooms or on parade grounds, but accompanying Iraqis "outside the wire." One, Lieutenant Colonel Terrence Crowe, bled to death on the streets of Tal Afar after an insurgent shot him in the femoral artery. Another, Sergeant First Class Paul Karpowich, spent several days conducting operations with his counterparts only to die when an Iraqi wearing a suicide vest bomb blew up the dining facility in Mosul. Three other members of the division were killed while performing advisory duties in parts of the country further to the south.

A transition team from the 98th trained the soldiers of what would be my Iraqi battalion. These Americans had taken over from the original cadre, who came from the Australian Army. The Iraqis completed their initial instruction at the end of 2004. Six months later, the soldiers of the battalion moved to a combat outpost

2 See Marine Corps 2005.

at Ba'aj, where they did convoys, patrols, and searches under the guidance of the 98th Division transition team and an American cavalry troop.[3]

The team from the next reserve institutional training division, the 80th, arrived a few months later to join the battalion as it participated in the September fight to clear insurgents from the nearby city of Tal Afar. They returned to Ba'aj afterwards, spending much of the winter of 2005-2006 helping to secure polling places during the constitutional referendum and national elections. Coalition casualties reached a low point that spring. The major I was replacing said in our emails that conditions were austere but that the battalion had made progress. As the Iraqis there and elsewhere across western Nineveh Province became more proficient, the US Army began turning over territory to the Iraqi Army and withdrawing to larger bases.[4]

The improving situation meant alterations to the team structure. All of my drill sergeants received assignments elsewhere by the time I was mobilized in April 2006. Their replacements included officers and senior sergeants with specialties in logistics, intelligence, transportation, and communications. These men had the knowledge to help the Iraqi battalion develop higher-level command and staff functions. I was disappointed that the US Army had upended months of team-building efforts with this last-minute change, but I was happy to have the expertise of the newcomers. As my reorganized team began training together that May at Fort Hood, Texas, the idea that the mission was primarily one of teaching remained implicit even though the late additions now made us capable of calling for close air support and medical evacuation.

Conditions across Iraq began to deteriorate after the February 2006 bombing of the golden-domed Al-Askari Mosque in Samarra touched off a civil war between Shia and Sunni Arabs (Worth 2006). In March, five US soldiers in Mahmudiya raped a fourteen-year-old Iraqi girl then murdered her and her family (Wong 2006a). Insurgents retaliated by kidnapping two Americans from a guard post, killing them, and dumping their mutilated bodies (Wong 2006b).

These events took place well to the south of Ba'aj, but they had profound implications for my team. The Coalition withdrawal accelerated as US units were moved from Nineveh Province to hot spots in Baghdad and Ramadi. The shifts left cities like Mosul and Tal Afar with less support and smaller towns like Ba'aj with almost none. The abductions raised further concerns about the security of transition teams like mine.

The American maneuver commanders decided that we could not "embed" with our counterparts. We would instead live about 20 miles to the north at a combat outpost near the base of Sinjar Mountain. We would share this home with the

3 For additional information about the 98th Division in northern Iraq, see Clay 2007 and Shipman 2006. During 2004-2005, Shipman was an advisor to the division of which the battalion in Ba'aj was a part.

4 For additional information about the 80th Division in northern Iraq, see Ciesinski 2006, Meier 2006, and Little 2007. During 2005-2006, these men belonged to the transition team that advised the brigade of which the battalion in Ba'aj was a part.

Iraqi brigade headquarters, an Iraqi battalion, a US cavalry troop, a Special Forces detachment, other transition teams, and encampments of British and Turkish civilian contractors. The change required my men and me to make three to five convoys per week on a route that grew progressively more dangerous the farther south we went.

The predominance of a religious minority called the Yezidi contributed to the relative peacefulness of the Sinjar district. These people had long been persecuted by Arab Muslims, so they tended to support the Coalition Forces. They received assistance from Kurds seeking to incorporate the land into an autonomous region that lay north of the mountain. Although some Yezidis resented this intrusion, hope for a better future manifested itself in the freshly-paved roads, new electric lines, and renovated businesses that had begun to appear by 2006. The villages of Jazeera, Qahtaniya, and Ananiya marked the southernmost boundary of Yezidi territory and of these improvements.[5]

By contrast, the inhabitants of Ba'aj and its environs were Sunni Arabs. Even Saddam Hussein had found them difficult to control. They regarded the Yezidis as devil-worshipers, and they used pipe bombs to drive away the few Shia refugees who had dared to settle there. They fought with each other, too. The area's two main Sunni tribes—the Shammar and the Khartouni—engaged in a sometimes bloody struggle for power.[6]

Many people here used tribal and family connections in Syria to smuggle fuel, cigarettes, livestock, and other contraband across the nearby international border. A sparsely-populated desert plain opens beyond Ba'aj. It contains deep wadis in which to hide and primitive roads connecting to cities along the Tigris River like Mosul and Tikrit as well as ones along the Euphrates River like Ramadi and Fallujah. Al Qaeda terrorists built upon this network to import weapons, ammunition, explosives, foreign fighters, and automobiles for car bombs (Felter and Fishman 2007).

Units of the US 101st Airborne Division under Major General David Petraeus had occupied Ba'aj after the initial invasion of 2003, but the American presence quickly dwindled.[7] During late 2004, lawless elements destroyed the police station, looted the bank, and drove away the mayor. The US 3rd Armored Cavalry Regiment under Colonel H.R. McMaster reestablished control during the summer of 2005. Despite this effort, the area continued to serve as a waypoint for smugglers, a

5 Maisel 2008 provides an excellent overview of the Yezidi.

6 Ivorybill 2007 provides an excellent overview of the political dynamics of western Nineveh Province.

7 See Arnold 2005. Arnold commanded the battalion that oversaw western Nineveh Province. Major General Petraeus concentrated his efforts during this time on pacifying Mosul, which degenerated into chaos upon his departure.

training ground for incoming terrorists, and a haven to which insurgents in other parts of Iraq could withdraw.[8]

The dire poverty of Ba'aj did not help matters. The majority of men did not have jobs and relied upon government food rations to feed their families. The streets were filled with puddles of raw sewage, piles of garbage, and wandering cattle. Water from local wells contained too much salt and other minerals for humans to drink it, so municipal trucks brought water from Sinjar. Electricity flowed from the power station for only four hours per day. Although some places had generators, shortages of diesel fuel rendered them inoperable much of the time. The maternity ward at the hospital swarmed with flies. Typhoid was a common disease.

Iraqi soldiers stationed at Ba'aj operated from an encampment set up in the ruins of the town granary. The multi-story, Soviet-built silos were the tallest structures around and could be seen for miles across the desert. The Iraqi Army used the towers as an observation post and to mount radio antenna and machine guns. The insurgents used them as a reference points to aim mortars and rockets. Below, the tents in which the Iraqis lived had earthen barriers on the sides but no overhead protection. The only available bunkers lay scattered a few hundred yards away on the empty half of the compound that American forces had occupied.

A US cavalry troop, Special Forces detachment, and transition team had shared quarters with the Iraqi battalion when it arrived in June 2005 (Olson 2007: 87). The Americans gradually withdrew as the Iraqis demonstrated their proficiency. The last US forces departed in May 2006. The next morning, a barrage of mortars caused a fire that burned down a tent and killed two of the Iraqi battalion's five company commanders. Regardless, the Americans officially turned over control of the sector with great fanfare and celebration when my team arrived in July.

Arrival: The Iraqi Army in Ba'aj

Although the members of my team and I did not really know what to expect, the optimism of the handover ceremony reinforced the idea that we would be teachers. In fact, I consciously intended to engage with my Iraqi counterparts in the same way that I normally establish authority and build rapport with a new class of students.

Circumstances did not permit such an approach. The Iraqi lieutenant colonel who commanded my battalion was fired a week or two after the team arrived. His replacement did not arrive for another month. The S2 intelligence officer quit, too.

---

8 See Yingling 2006 and Olson 2007: 85. In his 2006 State of the Union address, President George W. Bush would praise McMaster for his counterinsurgency tactics during this campaign. Yingling served on the regimental staff of the Third Armored Cavalry. He later gained notoriety for writing an astonishingly candid article titled "A Failure in Generalship" (see Yingling 2007). Olson was an intelligence captain for one of the squadrons within the Third ACR.

The new one arrived a month later and then immediately departed on emergency leave to arrange ransom with kidnappers who had abducted his father. The S3 operations officer missed three of our first four months to attend a staff school in Germany. The S1 personnel and S4 logistics officers lived two hour's drive away on another compound and visited once or twice a month. The only officer with whom we engaged regularly during the first weeks was a Kurdish major who served as the executive officer.

Even without this turmoil, the normal Iraqi Army leave policy militated against the kind of consistency needed for effective mentoring. Every officer and soldier was authorized ten days of leave per month. Iraq no longer had a banking system, which meant members of the armed forces received their pay in cash and needed time to carry the money home to their families. Ten days often morphed into fifteen, however, as some men worked out deals whereby only one at a time would work the same job. Fearing retribution from the superior, the junior of such "half-on/half-off" pairs would almost always resist any of our attempts to teach them anything new or make any substantive changes.

Their concerns were justified. On one occasion when the S2 officer was absent, my transition team intelligence officer successfully persuaded the senior sergeant of the section to take the initiative and conduct surveillance on his own. His efforts contributed to the capture of at least one insurgent and a weapons cache. Upon returning from leave, the Iraqi S2 officer fired the sergeant for not obtaining permission to act.

Meanwhile, the level of violence in our area escalated. Insurgents fired a rocket at the base during our first visit after the old team went home. It missed. They demonstrated better accuracy with mortars, which they lobbed regularly, usually in the morning or evening when Americans like us were not there.

Once during Ramadan, they hit two Iraqi soldiers while shelling the dining facility at sundown as hungry privates lined up to eat after a day of fasting. One had a piece of shrapnel deep in his elbow. The other had a spray of fragments buried shallowly up his back from his waist to his shoulders. The Iraqi lieutenant in charge of the aid station had locked up the medicine cabinet and gone on leave for two weeks with the keys, so his assistant dabbed antiseptic on the wounds of the unlucky pair without removing the fragments.

Insurgents placed numerous IEDs along the main route from Sinjar through Ba'aj. A bomb buried in the road shoulder rendered the mayor homebound until a surgeon in Damascus removed the ball bearings from his neck. An IED hidden under a sheep carcass on the pavement left an Iraqi soldier with minor head wounds. The S2 section's Land Rover was hit twice, but the vehicle's armor protected everyone inside from serious injury. A patrol from my battalion discovered another device before it exploded. The soldiers proudly carried it onto the compound and brought it into the battalion commander's tent before depositing it—electric blasting caps still in place—between the Iraqi colonel and me. Thankfully, they had at least disabled the detonator switch.

Law enforcement authorities provided little help. Terrorist gunmen ambushed and killed four policemen. They kidnapped and murdered a fifth. More than 30 Iraqi policemen from Ba'aj posted resignations on the doors of their mosques after receiving death threats. Forty others simply quit. Insurgents lobbed hand grenades at the police station and on one occasion blasted a hole in its second-story wall with a rocket-propelled grenade.

Civilians who sided with the government faced intimidation or worse. The head of the district council narrowly escaped an ambush on his way to a security coordination meeting. In October, three Yezidis who worked for the food contractor were stopped on the road between Ba'aj and Ananiya on their way home. Gunmen sprayed automatic weapons fire into the front cab of their small pickup truck. All three died. One of them was the sergeant major's brother.

Members of the Iraqi Army faced similar peril. In October, drive-by shooters hit one soldier in the head and a second in the back. Both miraculously survived. In November, terrorists fired bullets and mortars at a guard post and then detonated a suicide car bomb when the quick reaction force tried to pursue them. Three soldiers died in the blast. In December, an insurgent shot a private who had been standing guard. Wounded in the neck with only a shred of his spinal cord intact, the young man died of suffocation as an American helicopter flew him to Mosul.

More than a hundred Iraqi soldiers deserted from the battalion during the summer and fall of 2006, but not necessarily because of the danger. Leaving the armed forces before the end of an enlistment was not a crime, and conditions at the compound in Ba'aj were miserable. A shortage of fuel limited operational time for the generator that provided electricity and air conditioning. The corrupt civilian life support contractor failed to meet his obligations for supplying food, providing water, cleaning latrines, maintaining showers, collecting garbage, or making basic repairs to the infrastructure.

In addition, many soldiers came from villages in southern Iraq. Traveling hundreds of miles to take leave cost privates most of their pay, and they faced danger during transit. Several were burned by a firebomb. Two were imprisoned. At least one was shot to death. After returning late from leave, the second battalion commander told a harrowing story about being kidnapped at a false checkpoint, escaping, and walking barefoot for several miles until he reached safety.

Ordinary soldiers who returned late were assumed to have quit and removed from the payroll. The Iraqi Army did not have a system in place for reinstating them. To compound the problems, its system for sending new replacements had broken. As the present-for-duty size of the battalion dwindled, the workload increased for the soldiers who remained. They conducted patrols, performed guard duty, and conducted other missions with little or no time to rest. Exhaustion in turn led more desertions.

Our sister transition teams helped us stem the tide by persuading the Iraqi brigade commander to send a platoon from another battalion. This augmentation allowed soldiers from Ba'aj to come to Sinjar for a week. There, they had the opportunity to rest in a safer, less pressure-filled environment. They also underwent additional training. Medics from various units taught them first aid. Green Berets from a Special Forces team helped them with vehicle search techniques and rifle marksmanship. Field artillerymen from several transition teams gave instruction on how to navigate using hand-held global positioning devices. The respite boosted the morale of the Iraqi soldiers as well as that of their frustrated teachers.

The flood of desertions subsided. The effect of the program is difficult to measure, however, because the Iraqi higher headquarters somehow fixed its personnel problems and began sending replacements by the end of 2006. The overall number of attacks dropped after the battalion captured a second cache of weapons and explosives. Most significantly, a third commander took charge.

The second commander had been an Arab who did not get along well with the Kurdish officers who predominated at brigade and division. He was fired ostensibly for his handling of the November car bomb incident. More likely, he lost his job because he replaced the corrupt life support contractor, who apparently had high-level connections among the Kurds.

The third commander belonged to the same tribe as the brigade and division commanders. He was nevertheless well-qualified for the job. His previous experience included having led a battalion of Kurdish militia during the 2003 battle for Mosul. More recently, he had commanded the Iraqi Army military police at the division headquarters. Although he rehired the old life support contractor and took bribes from him, he was a very effective leader. He aggressively pursued the insurgents, and he cared about his soldiers. He consistently supported my team's efforts to work with the battalion staff.

Training Iraqis

Our instructional methods varied. We used a "crawl, walk, run approach" to teach our staff counterparts how to give their boss formal weekly updates. Our underlying goals were to encourage the Iraqis to think over the long term instead of reacting to immediate events, shine a spotlight on underperforming staff officers, and stimulate the exercise of administrative and logistical systems at the brigade and division levels. We developed a weekly agenda that contained statistical data like desertion rates, numbers of vehicles undergoing repair, and the amount of available fuel.

For the first meeting, Americans took the lead in putting the information together and presenting it to the Iraqi commander. The second time, Americans helped the Iraqi staff officers to prepare and then do the briefing. The third and subsequent times, the Iraqis did everything themselves while Americans monitored and assessed. Our Iraqi commander ensured that all of the staff participated, including a balky intelligence officer who viewed such briefings as beneath his station.

Unlike the Iraqi military, where officers customarily give all the classes, the US armed forces rely heavily upon noncommissioned officers as teachers. The ones on my team set a superb example. The team intelligence sergeant, who had experience as a civilian police officer, taught the soldiers of the Iraqi S2 section the correct way to take fingerprints. The communications sergeant, a Reservist who had served previously with a Special Forces A team, showed the Iraqi radio operators how to assemble and erect a directional antenna. He additionally helped them to establish a set format for reporting emergencies. The fire support sergeant, who had spent many years with active duty mechanized infantry units and who understood the subtleties of vehicle maintenance, mentored Iraqi drivers when the battalion received its first Humvees.

The medical sergeant, who came from the Regular Army, gave classes on field sanitation while arranging to obtain a floor, refrigerator, and air conditioner for the Iraqi aid station. He also worked in concert with NCOs from other US units to offer a weeklong Combat Lifesaver Course for Iraqi soldiers from across the brigade. The instructors followed the standard US Army curriculum, which consisted of lectures, practical exercises, and rigorous certification testing that included hooking an intravenous line to a live patient and administering a pint of fluid. Scores of Iraqis completed the program successfully.

Conferences provided another way for people from different units to learn in a large group setting. Leaders of my battalion's headquarters service company attended one such event at their division. There, with representatives from other Iraqi units across the province, they heard briefings about procedures for ordering parts, repairing vehicles, and other administrative-logistical functions. My team's Headquarters and Service Company advisor—a Reserve major from the Quartermaster Corps whose impressive knowledge, resourcefulness, and interpersonal skills made him a tremendous asset—went with them.

Sometimes outside instructors came to the battalion. For example, a logistical training team spent several days at Ba'aj. The American sergeants of this group included a cook who taught operation of the mobile kitchen trailer and mechanics who gave classes on fixing vehicles. Although the instructors had good intentions and great expertise, they added limited short-term value. The Iraqi battalion never went on any missions that required a field kitchen. Furthermore, the American civilian maintenance contractor who supported the division refused to distribute parts. He had a vested financial interest in requiring that all broken vehicles be evacuated to the division for minor work that could easily have been accomplished at the battalion level.

For several months during the last phase of our tour, we had working with us a section of US cavalry consisting of two vehicles and ten men. We were glad for their presence because the maneuver commander of our area withdrew the US cavalry troop from Sinjar in January and required that the transition teams move to quarters in Tal Afar. Our commute tripled in length as a result.[9] Besides providing extra security on convoys and at Ba'aj, the cavalrymen taught Iraqi soldiers individual skills such as rifle marksmanship, room clearing, and land navigation.

9 See Farquhar 2008. During 2007-2008, Farquhar led the transition team that advised the brigade of which the battalion in Ba'aj was a part.

Two men on my team had civilian jobs in academia that carried over well into the military environment. The American intelligence officer was director of instructional technology at a Midwestern college. He created a database program for the Iraqi S2 to track suspects and detainees. With the assistance of the team's interpreters and a keyboard that typed Arabic characters, he made the spreadsheets user-friendly for both Americans and Iraqis. Frequently, he used his knowledge of computer viruses to stop infections caused by Iraqis surfing for pornography while on duty in the tactical operations center.

The table of organization for a transition team did not include an S1 personnel specialist even after the restructuring of early 2006. The omission was rather odd given the terrible problems the Iraqi Army had with its administrative systems. Fortunately, one of the captains on my team worked in the human resources and financial aid offices of a southeastern university. He did a great job helping his counterparts track problems with pay, promotions, and basic accountability. He also supervised the team's interpreters by managing their pay and leave and helping to resolve issues with the American defense contractor who oversaw them. In addition, he handled the various discretionary funds the team received through the US military. Creating an unofficial position for a personnel advisor left the team without one for operations, so I substituted the field artillery captain who had been assigned from the Regular Army as the fire support coordinator. I gave him the duty of developing plans for the near, short, and long range. He very capably set up a training schedule that took into account the Iraqi leave system. Better still, he had the soul of a teacher. He clearly enjoyed watching people learn, and he constantly looked for new ways to foster that process.

His most ambitious idea took an integrated approach. He sought to create a "model platoon." The unit would have hand-picked leaders and the best soldiers. They would receive intensive individual and collective training. The rest of the transition team would work with the Iraqi staff to ensure that the soldiers in the platoon had all of their authorized equipment and that their personnel records were in order. The result would be an organization that would show the rest of the battalion at multiple levels "what right looks like." The Iraqi battalion commander agreed to the plan and had selected the platoon leader and senior sergeant by the time my team departed in June 2007.

The only person on my team who did not do much teaching—aside from a senior sergeant who was evacuated back to the United States early during our tour—was a young Transportation Corps captain from the Regular Army. He had responsibility for working with the life support contractor. He took a bullying, hard-line approach with the corrupt Kurd and played a key role in getting him fired by the second Iraqi battalion commander. Unfortunately, circumstances beyond our control led to the rehiring of the contractor several months later. The Kurd treated the captain with contempt thereafter, but the service did not return to its abysmal levels of old. Although I praised him in his efficiency report, and although he became a hero among the Iraqi privates, he became dispirited and did not contribute much to the team during the last part of our tour. The "school of hard knocks" can be painful for both student and teacher, but sometimes this method is necessary regardless of the alienation it can cause. That was the case here.

Just as the US Army resembled the American education system in its application of various pedagogical techniques, so did the military institution rely upon report cards. Advisory teams across Iraq used a format called the Transition Readiness Assessment. Every month, we assigned ratings of "Red," "Amber," or "Green" to the commanders and staff. We took the TRA seriously. All members of the team participated in its preparation, and we shared the results with the Iraqi commander. We never gave inflated ratings, and we never received pressure to do so. Subsequent format modifications, however, did seem to make bad news less apparent.

We Americans learned as much from our counterparts as we taught them. We began to understand that Iraqi administrative and logistical systems often did not work because of corruption and ineptitude at higher levels. I came to admire the resourcefulness of officers who used some of the bribe money they received from the life support contractor to purchase civilian parts that they needed for the battalion's Humvees. They also hired a local civilian to modify Toyota starters so that they would fit on military vehicles. Amazingly, they bought 12.7 millimeter bullets on the black market for the heavy machine gun that the Iraqi division had distributed without ammunition.

As for me, my job consisted to a large extent of sitting in the commander's tent with a cup of tea. Gathering intelligence about the latest battalion operations and insurgent activity required patience and listening. Like professors, Iraqis can talk endlessly without reaching the point. The commander would often tell me that nothing of importance had occurred since the team's last visit to Ba'aj. Not until later in the conversation would he mention, sometimes in passing, an IED, shooting, or nighttime raid that most Westerners would consider of vital interest. Ferreting out the information with my counterpart required the same kind of long-term perseverance needed to shepherd a student through a seminar paper, build rapport with a class across a term, or push a new program through a faculty committee.

Some of the regular US Army officers I encountered lacked the requisite patience or willingness to listen. They demanded immediate results and wanted to take over at the slightest delay. They expected advisors to issue orders to their counterparts. Neither I nor anyone on my team had that authority. My Iraqi battalion commander technically outranked me even after I received promotion to lieutenant colonel midway through the tour. He answered to his higher Iraqi headquarters, not me. When a US Army commander wanted to conduct a joint operation, I went to the Iraqi brigade commander. I had a good relationship with the one-star general, and he usually cooperated if I waited for him to finish his nap. My counterpart in Ba'aj knew about these meetings and seemed to appreciate my efforts to obtain permission from his boss, who did not like subordinates to take initiative.

Besides cultural awareness, writing prose was a skill from academia that helped me in my military duties. Our higher headquarters expected us to produce unbelievable amounts of paperwork. By the end of our tour, the requirement that we submit a situation report every 72 hours metastasized to 24. Each bullet we fired generated additional forms. Serious incidents meant collecting sworn statements from eyewitnesses. We also had to do efficiency reports and submit awards, both of which underwent endless revisions.

Handling college students who angled or begged for grades at my university gave me good practice for working with Iraqis. My counterparts incessantly asked me to buy things for them. They knew we had a pot of money intended for their benefit. Although they struggled to make a training schedule, they could produce a shopping list at a moment's notice. They particularly desired a color printer in order to make PowerPoint slides to send to their higher headquarters. I denied that request and many others as tactfully as I could.

As with students, I guided my counterparts toward things I hoped would have lasting value. We paid for someone to pour a concrete slab so mechanics could work on vehicles without lying in the dust or mud. We purchased bricks, mortar, and rebar to renovate the detention facility. We hired a civilian crane operator to move concrete mortar bunkers from the old, empty American half of the compound to where Iraqi soldiers could more quickly find cover. We mail-ordered a metal detector to find weapons caches and booby traps.

Education of an Educator

Prior to deployment, I made preparations to leverage my historian's knowledge of the American Civil War. I carried overseas with me a copy of an 1865 photograph depicting the rubble of Columbia, South Carolina, after Union troops under General William Sherman had passed through. I intended to draw comparisons between Iraq and the United States. My reasoning went: "Both of our countries fought a civil war. Both witnessed death and destruction. We Americans recovered. So can you." I bounced the idea off my interpreters. They were not impressed. "Baghdad looked a lot worse than that," said one disdainfully after looking at the picture. I dropped the idea.

Our interpreters served as cultural tour guides as well as translators. They went out of their way to explain nuances and shades of meaning. They were astute observers who tried hard to help us avoid misunderstandings. For example, one of them stopped a captain on my team from learning the hard way that the battalion executive officer would be gravely insulted by the flatulence some American males find amusing.

The interpreters also played up my status as a professor and doctor of philosophy. Such positions and degrees have prestige in Iraqi society, which is much more class-oriented than the American. Although I had completed the US Army Ranger Course and had spent two years as a Regular officer with an infantry battalion on the Korean Demilitarized Zone during the late 1980s and early 1990s, I had not participated in the invasion of Panama or the first war with Iraq. I had not served in a frontline unit for fifteen years. Emphasizing my civilian position and education hardly compensated for a lack of combat and operational experience, but I believed in using every resource at my disposal.

By contrast, my counterparts had lived on a battlefield for years and did not have the luxury of going home after twelve months. The first two commanders had made careers in the old Iraqi Army. They had fought in the war with Iran during the 1980s and against the United States during 1991 and 2003. Both had attended the elite Iraqi staff college. The third commander and the executive officer had belonged to the Kurdish militia known as the *Peshmerga* or "those who face death" and had waged a guerrilla campaign against Saddam Hussein during the 1990s. The operations officer, an Arab, had commanded a T-72 tank as a member of the Republican Guard and had battled the US military as recently as 2003. They had seen more fighting than most American soldiers, let alone me.

Academic life had further left me with habits of mind that did not translate well to military culture or combat. Commanding men under fluid, highly-stressful circumstances is very different from managing a like number of pupils within a microcosm bounded by syllabi, books, and the professor's whim. Life-threatening situations are not appropriate moments for leisurely contemplation or debate. They require quick, sometimes instinctive action and decisiveness.

Events gave me opportunity to practice. Not living with our battalion meant the team had to make frequent convoys. The desert wadis and weight of the armor on our Humvees limited our routes to a single road, which made avoiding IEDs and ambushes more difficult. Fortunately for us, the insurgents in this area did not go out of their way to target Coalition Forces as was the case in more dangerous locations like Baghdad or Fallujah. They either feared our heavier weapons or did not want to provoke the United States into shifting resources. On several occasions, they ambushed Iraqis only a few minutes after we had passed through.

The most consistent stress came from the warning shots we fired at civilian drivers who failed to obey the rule to pull over and stop for military convoys. The danger of car bombs made this policy necessary, but that did not make following it any easier.

Conditions changed when we went left the Sinjar-Ba'aj corridor. I came one decision away from getting blown up by an IED going into Mosul, and a sniper grazed my turret gunner on the way out. The bullet left two holes in his uniform and a circle of burned, bruised skin around the nick on his upper arm. A few weeks later, he spotted a possible IED in one of the wadis that intersect the road between Tal Afar and Sinjar. The roadside bomb blew up after he ventured too close while attempting confirmation. Miraculously, he suffered nothing besides a bloody lip and a mild concussion. This gunner, who was also the communications sergeant, had received minor burns to his face from an IED during a previous tour of duty in Iraq. He was one of the luckiest, bravest men I have met.

The latter road became of still greater concern after we moved from Sinjar to Tal Afar in January 2007. We traveled it on a regular basis for the remainder of our tour. To make matters worse, a group of insurgents decided to target the most desolate stretch. They started in late February by blowing up the culvert spanning the deepest wadi in the area. My team and I were the last Americans to pass. About an hour after we crossed, twelve sacks of explosives stacked underneath were detonated.

A few days later, as we approached the same site, a projectile splattered an unknown red substance on my vehicle. We never determined the identity of the fibrous goo, although we did find a circular crater next to the road that appeared to be the point from which it was launched. No one suffered any ill effects, and the expert at our base in Tal Afar concluded that it was not a chemical or biological weapon. Perhaps it was a testing or marking device. We jokingly called it the "Ketchup IED," but we all began to feel much more nervous about our long commute.

We had good reason for concern. Over the next two months, insurgents began laying more lethal IEDs nearby. The tall grass that had grown during the wet season made detection much harder. Several Iraqi soldiers and policemen lost their lives and limbs to roadside bombs hidden by the weeds. In another incident, secondary devices wounded an American lieutenant and started a fire that burned a US wrecker truck beyond repair. Although the Iraqis established new guard posts in the area, the attacks ultimately stopped as a result of an American sniper team and Coalition Forces aircraft.

Conditions meanwhile grew worse for my counterparts in Ba'aj as insurgents shifted northward from Anbar Province in response to the American surge and the Sunni Awakening. The attacks grew in frequency, intensity, and sophistication. Four days after the start of the New Year, a barrage of twenty 120 millimeter mortar rounds destroyed the S2 tent, perforated a shower trailer, and wounded the executive officer's hand. The destruction could easily have been worse.

In February, gunmen waiting along the road from Sinjar wounded the life support contractor in an ambush. Later that month, a suicide vest bomber killed five civilian workers who were building a new compound for the Iraqi Army across the street from the granary. Five more were wounded. In early April, soldiers manning the northern guard post successfully defended themselves from a coordinated assault involving dozens of well-armed attackers.

The Iraqis of my battalion worked hard to expand their influence over Ba'aj and the surrounding villages. They manned static guard posts to the north and south along the main road and sent roving counter-mortar patrols to the east and west. A quick reaction force remained on call at the combat outpost, which was ringed by barbed wire, an earthen barrier, and guard towers. Two men kept a 24 hour watch atop the grain elevator. Their 12.7 millimeter heavy machine gun, which they mounted in February, could fire for more than a mile in every direction.

When the battalion received intelligence either from its own S2 section or its parent brigade, the commander would sometimes lead platoon-sized raids.

These usually took place late at night when the transition team was not present. I suspected that the operations sometimes had as much to do with expanding Kurdish influence as they did with combating the insurgency, but I had no proof and could do little to prevent such activities. I did make certain that someone from the team checked on any detainees to prevent them from being abused.

The primary mission the battalion conducted as a whole was the cordon and search. These usually took place as part of a larger division or brigade sweep. The battalion would have two or three villages as objectives. Usually one company would impose the cordon, which meant surrounding the target and preventing anyone or anything from going in or out. Another company would clear each house.

Twice the battalion did a variation on this mission in which one company cordoned the village and another brought a truck into the center to distribute food. The intent was to foster goodwill toward the national government, demonstrate the reach of the Iraqi Army, and undermine the influence of the insurgents.

My transition team went on more than fifteen battalion-or-higher-level operations. We had the specific tasks of calling for air support or medical evacuation if needed and serving as liaisons to any American units involved with the operation. We monitored the Iraqis to determine the effectiveness of the cordon and the thoroughness of the search. I normally stayed near the battalion commander.

Teaching took a lower priority on these missions. We would have liked to have walked the staff and company commanders through the standard American planning system. Although the third Iraqi battalion commander supported almost every initiative we proposed, he resisted this one. Instead, he told his subordinates to prepare for a generic cordon and search. They did not learn the details until the last minute. Sometimes, he did not tell anyone the location of the village until the battalion was en route or had arrived. He withheld information primarily out of concern for security.

I disagreed at first, but as time passed I came to suspect that insurgents had infiltrated the battalion. Some of the later mortar attacks suggested someone inside the compound was providing information about the locations of precise targets and the arrival of certain visitors. The lesson was driven home over a year after I returned home. On November 25, 2008 an Iraqi soldier assigned to the battalion opened fire on the senior sergeant of the team who was then working in Ba'aj. Master Sergeant Anthony Davis died in the attack as did a visiting Marine, Captain Warren A. Frank. They had been distributing food to civilians (Robertson and Farrell 2009).

The battalion-level missions did not seem especially dangerous at the time. The soldiers almost never found anything of importance because word leaked out in time for any insurgents to disappear. I came to enjoy having the opportunity to meet local civilians, often over breakfast. In fact, I met a considerable number of teachers on the occasions in which the battalion command post was located in a village school building. Classes were normally cancelled, which left educators with little else to do during the interim but drink tea and talk. I learned much about the condition of the facilities and the bleak outlook most Iraqi children faced. Nevertheless, I was surprised to hear many of the teachers express nostalgia for the Saddam Hussein regime and the comfortable, regular salaries they had enjoyed.

One of the two times my Iraqi battalion did find anything of significance was during November when it participated in a brigade-level sweep of Ba'aj. Among the detainees was a man who taught physical education and history at the city school for boys. (Apparently, coaches gravitate towards social studies in Iraq, too.) He also belonged to the Iraqi branch of Al Qaeda and admitted his role in several recent incidents. He and the Iraqi battalion executive officer had a discussion about terrorism that could have taken place in a faculty lounge had not the conversation been in Arabic and had not one of the participants been wearing flex cuffs. The teacher quite calmly stated that he would do his utmost to kill me and any other available Americans if he were released. I had visited the detention facility after an exhaustingly long day to make certain that my counterparts did not abuse their prisoners. His ingratitude was chilling.

Experiences such as these dissuaded me from collecting supplies or making the kinds of efforts to help Iraqi schools that Coalition Forces soldiers sometimes organized elsewhere. Two Sunni villages near Ba'aj remained hotbeds of insurgency despite projects to renovate educational facilities. At a third, news that an American Civil Affairs unit planned to make improvements led local civilians to strip the existing building of doors, window frames, lights, and plumbing. They even tore holes in the reinforced concrete walls to rip out the electrical wiring. After the United States fixed the girls' school in Ba'aj, the headmistress sent students home rather than allow a dedication ceremony to take place. In addition, someone on the faculty or staff appeared to have encouraged six to ten-year-old girls to defecate all over the floor of the bathroom in anticipation of our arrival.

The only other instance in which my counterparts found anything during a large-scale, planned cordon and search was in May when we searched a remote village in a part of the desert that no Coalition or Iraqi armed forces had visited for a long time. We caught a man frantically trying to make a call on his cell phone as we arrived. He clearly was trying to warn someone of our presence. Some solders from our battalion captured him and searched his house. He had two automatic rifles and hundreds of bullets in his possession. Iraqi families are only allowed one such rifle and 30 rounds of ammunition. According to the passport we found he had recently traveled to Jordan. Despite my recommendation that the civilian be detained, my commander decided to release him in order to foster good community relations. The detainee's father demonstrated his thanks by inviting us to share a lunch of flatbread, ewe's milk, and pan-fried sheep testicles. My stomach could handle anything by that point.

The most dangerous missions on which we participated were ones in which we moved to contact. Although the insurgents tended to attack when Coalition forces were not nearby, they became bolder by the middle of 2007. Several incidents occurred while we were present. In late April, insurgents mounted a 14.5 millimeter antiaircraft gun on the back of a truck and exchanged fire with a counter-mortar patrol and the 12.7 millimeter gun atop the granary. In early May, they launched a coordinated attack on the guard post at the southern entrance to Ba'aj using rocket-propelled grenades, hand grenades, and small arms. My team and I ended up in a four-way cross-fire. Six people died in the firefight, and nine were wounded. Later in May, insurgents shot four 82 millimeter mortar rounds into the combat outpost as a US cavalry troop arrived to conduct a joint cordon and search. Two members of my team and I were accompanying them as liaisons. On all three occasions, we pursued the enemy and coordinated a response by Coalition aircraft. We never caught our opponents despite our best efforts.

By the end of our tour, I felt we had earned the respect of our Iraqi commander, not just as teachers but as soldiers. He took comfort in our ability to call for air support and asked us to remain near the battalion whenever it went out on large operations. He thanked us for being the only Americans to visit Ba'aj with any frequency.

Unfortunately, the neglect of western Nineveh Province by Coalition forces would have tragic consequences. In August 2007, two months after we departed, the Yezidi villages of Jazeera and Qahtaniyah were destroyed by four massive suicide bombs carried on three trucks and a car. Estimates of the number of civilian men, women, and children killed ranged from three to seven hundred. More than a thousand people were wounded (Glanz 2007). According to one of my interpreters, who continued working for the team that replaced mine, the battalion from Ba'aj was the first military unit to respond to the blast.

The United States military apparently began paying closer attention to the area after this atrocity. The next month, a Special Forces raid on an insurgent camp in the desert near the Syrian border captured what came to be known as the Sinjar Papers. These consisted of four terabytes of data containing details about more than seven hundred foreign fighters who crossed the Syrian border in our area from 2006 to 2007 (Felter and Fishman 2007).

Could my team and our counterparts in Ba'aj have prevented the Yezidi bombings? Could we have interdicted more foreign fighters? Those are questions that linger in my mind. I doubt, however, that we could have done any differently given the state of the Iraqi Army at that time, the vastness of the desert terrain, and the preoccupation of the United States elsewhere. Our battalion became a more capable fighting force, able to sustain itself better and exert influence over a much larger space than it had upon our arrival. How much credit we as advisors should receive is anyone's guess. Much of the improvement came out of necessity and from within the "students" themselves. Those are feelings any teacher, uniformed or civilian, should understand.

PART II
Military Academies and Humanistic Inquiry

Chapter 4
Teaching Citizen Soldiers: Civic Rhetoric and the Intersections of Theory and Practice at the Virginia Military Institute

D. Alexis Hart

Background

Although I was not a stereotypical "military brat," I did grow up in a military family. I also grew up in an academic family. My father, a college math professor and an officer in the Navy, came from a long line of teachers and military officers himself—both his father and his grandfather had served in the Army and had later worked at colleges. By the time I was born, my dad had resigned from the active service but was still an officer in the Navy Reserves. Like most reservists, my dad "drilled" one weekend a month, but rather than serving the typical two weeks of active duty a year, he went on active duty for a full three months during the summer, teaching math classes to help prepare enlisted personnel who had been selected for commissioning programs for college. In addition, he served as a Blue and Gold Officer (recruiter and interviewer) for the Naval Academy and the Naval Reserve Officer Training Corps (ROTC), so the Navy was a constant presence in our house—as were academics. Along with his military and teaching careers, my dad was a community leader in a grassroots citizens' group advocating responsible use of geothermal energy and an elected official on the local city council and a statewide "league of cities." In these ways, my father showed me the real differences active citizenship and a commitment to public service could make—whether it was recruiting the next generation of Naval officers, preserving natural resources, or ensuring the city streetlights stayed on at night. He attributed much of his own motivation to work for the common good to the examples set by his father and grandfather, and he credited his military experience for helping him to become a more effective leader. Sensing that military service might be a suitable beginning for my own engagement in civic life as well, when it came time for me to go to college, I sought out an opportunity to commission, and I was fortunate enough to receive a four-year Navy ROTC scholarship.

As a college student and a Navy midshipman, I saw little overlap between my academic courses and my military responsibilities until the start of the first Gulf War in 1991. Suddenly media teams appeared on campus to interview those of us walking around the college quadrangle in uniform. My friends and classmates began to confront me with questions and even some accusations about being complicit in trading "blood for oil." Being thrust unexpectedly into the public eye this way, I quickly realized that I needed to be able to articulate measured and well-informed responses to the kinds of inquiries and indictments my friends, classmates, and the media were lobbing at me. I became acutely aware that my "personal" statements could potentially have an impact not just on me but on my peers, my chain of command, the American public, and others. It was at this time that I started to form a better sense of how my English major and my Anthropology minor were in fact preparing me for my future military career. Prior to Desert Storm, I had been inclined to brush off questions about what I was going to do with an English major in the Navy with a quick quip that the Navy would train me to do whatever job I was assigned. After the war began, it became clear to me how valuable studying the literary and cultural traditions of a number of different societies could be for my future career. My schoolwork, I realized, was making me more aware of my role as a global citizen and was allowing me to develop skills that would help to make me a more effective leader—skills such as "active listening, deliberation, engaging diverse perspectives, collaboration, creative problem solving, civility, ethical decision making, and information literacy" (Jacoby 2009b: x). I also began to understand how beneficial it can be for military officers to have a broad (and, yes, liberal) education in order to be able to assess the complexities of any conflict or "peacekeeping" action, to make decisions based upon contingent knowledge, and ultimately to accept responsibility for their decisions—just as Odysseus or Henry V had to do.

I once again found myself at an intersection of academic and military cultures immediately upon graduating from college and receiving my commission. My first active duty assignment was to volunteer in a community service program—one of several such local programs supported by the US military. For two months, I taught an introductory technical writing class to middle and high school students in a pre-engineering program designed to provide "special outreach to Hispanics, African-Americans, Native Americans and females who have been historically underrepresented in the scientific, technical, engineering, and mathematics career fields" (Prep-USA 2005). This experience solidified my belief in the value of civic engagement and service to others. In this program, I had the opportunity to influence my students both through the curriculum I was presenting and through my presence as a Navy officer. By helping them to hone their research and communication skills, I was providing them with a better chance to succeed in future courses, and I was giving them confidence in their aptitude for college. Because I taught in uniform, I also served as a role model, and I found that my students frequently asked me about my choice to serve in the military. I candidly told them of my inclination to contribute to the common good—one manifestation of which was my decision to volunteer to teach them.

My second duty assignment was as a student myself at the Navy Supply Corps School (NSCS), where I learned how to be a business manager for the Navy. My graduation from NSCS happened to coincide with the repeal of the Combat

Exclusion Law, which had prohibited women from being assigned to combat ships. Consequently, I was assigned to the USS Essex, an amphibious assault ship whose mission is to "transport, deploy, command and support all elements of a Marine landing force of over 2,000 troops during an assault by air and amphibious craft" (USS Essex 2009). During my three-year tour, I completed two six-month Western Pacific deployments as well as several shorter operations at sea, and I had the opportunity to go ashore in such varied ports as Mazatlán, Hong Kong, Singapore, Malaysia, Thailand, Oman, the United Arab Emirates, Kuwait, Bahrain, and both coasts of Australia—I really did "join the Navy and see the world." The experiences I had while doing business and going sightseeing in these international ports made me more keenly aware of the privileges I enjoyed as a United States citizen, reinforced my nascent sense of the importance of communicating effectively (both at work and at play, in uniform and out), and enhanced my understanding of the potential impact my public words and actions as well as the actions and words of the rest of the ship's crew could have. In addition, traveling around the globe helped to expand my limited awareness of how interdependent the nations of the world are. As I developed this awareness, I attempted to use my academic research and critical thinking skills to become more informed about the different customs and cultures of the ports in which we were anchoring. I also encouraged the sailors and Marines in my division and department to listen attentively to the "port briefs" shown over the ship's closed-circuit television prior to every port visit. I asked them to try to respect the cultures we were encountering and to strive to make responsible, civil, and ethical choices when ashore—whether that meant wearing long sleeves even in the heat or refraining from chewing gum in public. I realized that as a division officer I was unofficially serving as an educator as well, and I saw how important it was for me to act as a role model if I wanted the enlisted personnel to follow my advice. In other words, my actions had to match my words if I was to retain any credibility with those whom I was trying to teach and to lead.

I more explicitly sought out this dual responsibility as role model and educator in my final active duty assignment when I returned to the Navy Supply Corps School as an instructor. Here, in addition to teaching junior officers how to fill out standard forms properly and to balance complex budgets, I found myself emphasizing the overall importance of clear communication and promoting leadership by example. While I was teaching at NSCS, I began taking graduate classes in English at the University of Georgia. I had determined that the high operational tempo (and resulting time away from homeport) of the active Navy was incompatible with my personal desire to have a more stable family life, and so I decided to pursue another avenue of public service as a teacher. By that time, I had formed a more thorough understanding of the links between my own education and my military service. Having completed three years of shipboard duty, I felt even more strongly that my exposure to a variety of literary, historical, and cultural texts as an undergraduate English major had helped me to develop an ability to appreciate diverse viewpoints and to respect different customs. I also believed that the reading and writing I had done in my literature courses had refined my communication and critical thinking skills. Believing that my competence in these areas had made me a more effective leader and a more responsible global citizen, I resolved to pursue graduate studies in English as well, although I was uncertain about an area of specialty. Due to the vagaries of balancing a full-time job with enrolling in graduate classes, the second graduate course I ended up taking was a Composition Pedagogy course, and it was here that I found my specialization.

In the discipline of Rhetoric and Composition, I sensed immediately the intersections between my past experiences as a military officer and my future opportunities as a professor. Due in large part to my non-academic experience, I readily accepted the premise that "knowing effective and responsible ways to investigate and write about a subject in which one is interested is fundamental not only to academic writing but also to participation in a republican form of self-government" so therefore "learning to understand and observe the obligations embedded in academic writing may contribute … to the development of a student's moral character as a citizen" (Stotsky 1992: 798). I also quickly recognized that teaching writing had the potential to be "a high stakes enterprise with implications not only for students' academic and professional success—important as that is—but also for the health of participatory democracy" (George 2001: 97). In my travels around the world, I had the opportunity to see firsthand the differences between participatory democracies and other forms of governance and had come to realize how much I had taken my freedom to engage in open debate about political and social issues for granted. I had also seen promising ideas dismissed or overlooked and events gone awry because of poor communication. Serving in the Navy, I had found that I made the most positive impact on my commands and on our missions not through ensuring my books were balanced and my inventories were correct (although those were certainly important) but through writing strong performance evaluations and award recommendations for the enlisted personnel and junior officers under my supervision and through establishing respectful and ethical relations with the people with whom we did business—both at home and abroad. As a writing teacher, I decided, I could take advantage of the opportunity to help my students not only improve their communication skills but also ask them to consider how they might use their words and actions to participate in our democracy.

As I continued my graduate studies, I was particularly captivated by classical Western theories of civic rhetoric and by the pedagogy of the ancient Greek rhetorician Isocrates. Even in the Navy, I had never been content just to follow the standard operating procedures (SOPs) and had rarely hesitated to question the traditional ways of doing business. I was not one to be content with the attitude of "if it's not broke, why fix it?" I understood the need for checklists and SOPs, but was also open to making changes based on good suggestions or new information. Therefore, I appreciated the fact that Isocrates did not view writing simply as a technical competency (that is, learning to follow the rules of grammar and punctuation) or just as a way of conveying an already existing reality (that is, composing the "correct" responses to an instructor's prompt). Instead, his view of discourse mirrored my own practical experiences. Isocrates regarded speaking and writing as a means of generating, organizing, and circulating ideas so that one could make judgments in the face of limited knowledge and could engage others in the decision-making process. In Isocrates' pedagogy, I found a teaching philosophy that had as its principal goals those elements of leadership I had found to be most relevant to me as an officer in the Navy: working out my ideas, making informed judgments based on contingent facts, communicating my decisions effectively, and working with others to make a difference. In what follows, I will discuss the wax and wane of civic education in Western culture and in American higher education in particular, describe my Isocratean pedagogy, and then discuss why this pedagogy is particularly well-suited for the overlapping academic and military cultures at the Virginia Military Institute (VMI).

Civic Education: Ancient Precedents and Modern Lapses

As historian Barry Strauss observes, in ancient Athens "[t]here was little separation of civil and military ... To the ancients, war was too serious to leave to the military—or to the professors. Instead, war belonged to everyone. The whole city taught war" (2008: 53-4). Being involved in politics was also looked upon favorably in this early society "since it was implied that the individual was escaping from self-centeredness and devoting his abilities to the *vita activa*, for the common good" (Vickers 1988: 7). Not surprisingly, therefore, many Greek educators (including Isocrates) prided themselves on training leaders for both the political and military arenas and encouraged even those students who sought other career paths to become active and engaged citizens. In fact, preparing young people "to be citizens in a democracy was the principal aim of education" in ancient Athens and was also "the principal aim of education in the time of Thomas Jefferson when he talked about the founding of public schools," as Derek Bok, former president of Harvard University, pointed out in his 2001 address to the Institute for College Student Values. Unfortunately, as Bok also noted, "somehow in the intervening years that purpose has been lost" (2001). Writing teacher and rhetoric and composition historian Robert Connors attributes the declining focus on training young people for lives of democratic citizenship in higher education in general and writing courses in particular to the circumstances of the nineteenth century, when "the very culture of rhetoric, which had always informed Western education, turned from a public, civic orientation meant to prepare leaders of church and state toward a more privatized, interiorized, and even artistic orientation meant to aid in self-development or career preparation in bureaucratic organizations" (1997: 23). A 2004 study by the Association of American Colleges and Universities (AACU) supports Connors's assertion. According to the AACU's findings, "students overwhelmingly [believe] that college [is] a time to prepare for a job and take responsibility for themselves and their own obligations" (Jacoby 2009a: 23).

Another consequence of this nineteenth-century shift in focus in higher education from the public citizen to the private individual has been the prevailing attitude among even educated Americans that one should "abstain from all social action except the pursuit of [one's] individual economic goals in the market, and voting for candidates for public office" (Ohmann 1976: 88). By contrast, for Isocrates the notion of a "private individual" was highly distasteful, as he made clear in his criticism of Plato's and Socrates' ideal of the man of wisdom who retreated from contact with the community and declined to engage in public debates.

As American university curricula have drifted away from a focus on moral development and public service, educators have likewise begun to spend less time teaching students how to engage with civic issues. As a result, many college students have a limited sense of responsibility to each other, to those they have power over, or to those in positions of power over them. They also are not being adequately equipped with either the capacity or the desire to use their discourse to become engaged public citizens. While an encouraging number of young people did turn out to vote in the most recent presidential election, we need to remember that "voting is the most minimal form of citizenship, demanding nothing more than showing up and hopefully pulling a lever or connecting the lines" or, now, touching a screen (Glickman 2002: 7). My concern is that many young people are content with this most inactive type of political involvement (that is, voting) and have little aspiration to engage in more active processes of democratic decision-making. Unfortunately, due to the self- and career-oriented nature of our current education system, many of our students continue to see their private lives as largely unrelated to the life of the polis. Therefore, they often fail to recognize themselves as social and political agents who can and frequently do have an influence upon various communities. What our students frequently lack is an awareness that "the public arena, by its very nature, requires … the sacrifice of self-interests for the larger good" (Condit 1999: 309). Indeed, a 2006 study by the Ford Foundation found that "there is near consensus that an essential part of civic engagement is feeling responsible to be part of something beyond individual interests" (qtd. in Jacoby 2009a: 8). This lesson was one I learned quickly and experienced repeatedly in the military, especially once I got to the Essex, where one of our oft-repeated mantras was "Ship, shipmate, self." Consequently, I hope to bring to my students an awareness of the importance of occasionally sacrificing self-interest to contribute to a greater good.

Isocrates certainly did not believe that the purpose of higher education should be confined to the achievement of each individual's selfish ends. He thought instead that higher education should lead an individual to develop greater virtue, more extensive knowledge, and keener judgment, and he believed his students ought to put their education to use by contributing their abilities to the public good. Isocrates would have been distressed to see his educational ideal of training students to become engaged public citizens being replaced with the contemporary view that individuals have the right "to be left alone to pursue personal advancement" (Miller 1997: 187). Yet this attitude is one that is often held by our students. After all, they have "come of age in a culture of increased privatization and renewed individualism," and therefore "have been socialized to think of themselves as consumers and measure their accomplishments according to marketplace standards" (Longo and Shaffer 2009: 154-155). Isocrates would have chafed at the idea that an individual's vocational ambition could be equated to service to society by way of his or her economic contributions alone.

As Isocrates might have suspected, our students' tolerance of or failure to interrogate the role of the passive consumer can in fact lead them to withdraw from or neglect to engage in public arenas beyond the marketplace. It can also limit their ability to see themselves as active agents who can contribute meaningfully to their communities and who can create, define, challenge, or transform contested concepts of the "common good." In order to enable our students to participate more actively in the public sphere, like Isocrates, I try to give my students both "the ability to *critique* marginalizing representations, disadvantageous subject positions, and biased modes of legitimation" and "the ability to *compose* empowering representations, advantageous subject positions, and yet remain inside the scope of 'legitimate' discursive practices within any given institution" (McComiskey 2000: 116)—institutions such as the military services or a military college. I recognize that in order for America's undergraduates to be prepared "for lives of moral and civic responsibility," they need "not only to possess knowledge and intellectual capacities but also … must be willing to act for the common good" (Colby et al. 2003: 7). It is this willingness to look beyond one's own self-interest and to assume responsibility for others that Isocrates encouraged in his pedagogy, that I internalized early by watching the example of my father, that I attempted to enact in my military service, and that I now try to foster in my classrooms.

While I openly encourage my students to consider ways of setting aside their personal ambitions in order to use their discourses to engage actively as public citizens, I maintain an awareness that "the recovery of a strong participatory idea of citizenship should not be done at the cost of sacrificing individual liberty" (Mouffe 1991: 72). I also keep in mind the fact that I need to avoid creating "a false dichotomy between individual liberty and rights on one side versus civic activity and political community on the other" (75). Therefore, I try to encourage an ongoing discussion about competing interpretations of democratic principles such as equality, liberty, and the common good because I recognize that aside from the military, higher education is "one of the few public spaces left where students can learn the power of questioning authority, recover the ideals of engaged citizenship, reaffirm the importance of the public good, and expand their capacities to make a difference" (Henry Giroux, qtd. in Jacoby 2009b: x-xi). One way I often open up these conversations is by sharing my own "sea stories" with my students (such as the time I told my commanding officer that he was wrong to prohibit female officers from serving as admiral's attendants during a change-of-command ceremony) to show them that even in the military one person *can* choose not just to follow orders, *can* question authority, and *can* advance the common good. I realize that not all of them will have an opportunity to do so through military service, as I did. Therefore, along with providing a space for learning about and discussing what it might mean to be a citizen, I try to provide opportunities for my students to practice civic engagement by composing written and oral arguments about issues that matter not only to them but to others as well. In this way, I try to combine theory and practice in my courses, and, in doing so, I frequently return to the model of Isocrates.

My Isocratean Pedagogy

Because his school arose out of his desire to "compensate for the lack of enlightened statesmanship and mindful citizenship apparent throughout Hellas in the aftermath of the Peloponnesian War," Isocrates took great pains to link his teaching "to the articulation of wise governance and civic conscience" (J. Poulakos 1989: 317). At his academy (one of the first permanent schools of higher education in Western history), Isocrates strove to develop young men (and all his students *were* men) who would become generals and politicians, historians and lawmakers, teachers and democratic citizens, declaring that "those studies which will enable us to govern wisely both our own households and the commonwealth … should be the objects of our toil, of our study, and of our every act" (1928-1945: 285). His pedagogy therefore incorporated a high degree of patriotism, and he entreated his students not to be satisfied with voting alone, but urged them to be more actively involved in seeking the best interests of the state. Furthermore, he declared that the success of his students should not be measured simply by the pupils' skills as speakers and writers, but more importantly by their explicit performances of civic virtue. In fact, he claimed that those students who followed the "true precepts" of his program would be "helped more speedily toward honesty of character than towards facility in oratory" (21). Thus he hoped that even if his students had a limited ability "to make proper use of [discourse] at the appropriate time, to observe the right sentiment about [it] in each instance, and to set [it] forth in a finished phrase" (9-10) they would still make an honest effort to articulate publicly their versions of the "truth" and to present them as the basis for subsequent public debates and social actions. By asking my students to do the same, I hope with Isocrates that "those who most apply their minds to [the specific occasions of their discursive performances] and are able to discern the consequences which for the most part grow out of them, will most often meet these occasions in the right way" (184). I therefore encourage my students to approach every act of writing or speaking as an opportunity to engage with a real audience who has a genuine stake in the issue—whether their classmates, their teammates, the people back in their hometowns, or their political leaders. I also tell them that they must be willing to accept responsibility for the ultimate consequences of their texts—just as those of them who will give orders to troops, advice about stock options, or recommendations for building materials must accept the consequences of those decisions and actions.

In my classes, I first present the notion of public engagement through a lens of Isocratean rhetoric and then try to provide my students with opportunities to write and speak about the issues that impact their everyday lives and the lives of those around them. In this way, I hope to offer them an opportunity to enter conversations to which they can contribute meaningfully and to make assertions for which they can take responsibility. By encouraging them to think about how they can work to make a difference—whether by trying to convince the auxiliary services officer to improve the nutritional value of the mess hall food or by encouraging their local representatives to vote for or against a ballot measure—I try to urge my students not to be satisfied just with voting every couple of years as a way of participating in the democratic process. In this way, I also hope to get them to move beyond their individual interests in order to recognize their roles as public citizens.

Even though he was faced with problems similar to those we encounter today, including the decline of civic-minded public discourse and the elevation of privatized individual interests over broader community concerns, Isocrates took seriously his responsibility for training public citizens who could seize upon opportunities to create changes in the social, political, and material conditions of their various provisional communities through public discourse. As I mentioned earlier, Isocrates encouraged all of his students to consider becoming political or military leaders (much as VMI encourages all cadets to consider becoming military officers), but he knew that not all of them would wish to do so or be capable of doing so (nor do all VMI cadets choose to commission). Isocrates felt that for the latter his training was at the very least "capable of leading them on to self-improvement and to a greater degree of intelligence on many subjects" (15). Such traits, he believed, would give all of his students the basic competence to contribute thoughtfully to the public welfare as actively engaged citizens as well as the ability to act as models for their families and other citizens to emulate. I, too, seek to provide my students with these competencies and leadership traits.

Isocrates regarded as most capable to lead those who "regarded the care of the common interests not as a business, but as a public service" (25). It is this shift from a limited business- or job-oriented focus to a broader ethical commitment to the welfare of others that I push my own students to make. I believe that my students need "to see their 'opinions' as historical and social products (and not as their own private property), which can be critiqued and transformed" (Katz 1995: 214) and which can themselves be transformative. In other words, as members of an increasingly global society, I think my students must come to understand that there are various ways of telling "the truth" and that they are responsible to others for the "truths" they themselves tell. Therefore, I ask my students to view the texts they produce in my courses as rhetorical public performances, and I encourage them to deliver their texts to their intended audiences—whether in person, by mail, or through the Internet. I want them to see that their classroom assignments do not just have to be products that are assessed for a grade, but that their discourses can potentially result in influential public *actions*, and, furthermore, that these actions can produce serious and lasting social, material, and political *consequences* for themselves and for others. These are the lessons I learned most keenly during my shipboard experience—whether in the ways I gave instructions to the helmsman to change course, in how I corresponded with the vendors providing port services, in how I phrased a performance evaluation or an award citation, or in how I comported myself on liberty. I have found that by following Isocrates' example and regarding my students first as citizens (or, as "citizen soldiers" in the VMI context), I can often help them overcome their initial inclinations to think of the written and spoken texts they produce in my classes as having an effect only on themselves personally—their grades, their GPAs, their chances to get a desirable job or get into a highly regarded graduate school. And, by extension, I hope to make them aware of the ways in which they can continue to use their rhetorical performances to engage as citizens once they leave my classes and the academy.

My conception of what it means to be a "citizen" in the contemporary context relies heavily upon the definition of critical theorist Etienne Balibar, who claims that the citizen "properly speaking is neither the individual nor the collective, just as he [sic] is neither an exclusively public being nor a private being" (1991: 51). As Isocrates might say, a citizen ought to seek to be an active participant both in the commonwealth and in his or her household. Balibar acknowledges that a citizen can preserve his or her individual liberty while simultaneously being an active participant in public life. However, a citizen "is unthinkable as an 'isolated' individual," says Balibar, because it is his or her "active participation in politics" that brings about his or her existence as a citizen (51). As I help my students begin to consider the intersecting interests of individuals and the collective, they ideally come to apprehend that "[w]hatever the private goals motivating individuals to act might be, it is usually only once a public or the citizenry is persuaded to endorse and act upon communally shared goals that history moves forward (or backward) in significant ways" (Lucaites and Condit 1999: 4). As historian of rhetoric and teacher of writing Edward Corbett explains, "Training students to *know*, to *realize*, people other than themselves is what prepares them for membership in society" (1989: 200). Preparing my students "for membership in society" is thus my ultimate goal as a writing teacher and a college professor.

As I mentioned earlier, in the Navy I encouraged the sailors and Marines in my division to consider their responsibilities as members of a global society and as representatives of the United States overseas by asking them to view carefully the port briefs that gave quick cultural snapshots of the places we went ashore, to respect the customs they learned about, and to comport themselves in a way that would "reflect credit upon themselves and the Naval Service" (the standard text of Navy award citations). Similarly, in order to help the cadets in my classes to understand that different viewpoints exist even among the relatively homogeneous student population at VMI, I first ask them to consider their own personal histories and experiences and to examine how their opinions and beliefs have been affected by those experiences. I then ask them to consider how various events (such as the current war or the phenomenon of text-messaging) and public figures (such as the current president or pop-culture icon) have been socially constructed and how their perceptions of these events and individuals have been affected by diverse historical, ideological, technological, material, economic, and literary forces. I tell them that they should continue to cultivate this awareness if they wish to enter ably into public conversations, that is, if they wish to have their voices heard and acted upon in the provisional communities to which they belong—whether in the barracks, out in town, in their jobs, or in the larger world. Ideally, as my students begin to define themselves more self-consciously as citizens rather than as individuals isolated and even insulated from the public world, they also begin to realize that their ideas and goals do not originate and terminate solely within themselves but are enmeshed in and can have an effect upon larger communities and that they therefore are responsible to others in those communities.

As I try to cultivate their awareness of the forces that act upon them and their communities, I also alert my students to the fact that as citizens they paradoxically encompass "both the point of view of the State apparatus and that of the permanent revolution" (Balibar 1991: 55). In other words, citizens do not just follow the rules or obey orders without question. Instead, they are continually suspended between maintaining loyalty to the state and overturning institutionalized policies, sometimes favoring one side, sometimes the other, and deciding on a course of action based upon balancing the best interests of themselves as individuals and the best interests of the larger community. This discussion of citizenship as a tightrope act between abiding by the rules and questioning authority seems to resonate particularly well with the cadets at VMI, an institution with numerous time-honored policies that are distinctive from other colleges and universities and that the cadets constantly have to negotiate—as individuals, as members of the Corps of Cadets (the student body), and as part of the larger institution itself. Like Balibar, Isocrates believed that a citizen had to be able to weigh critically the various options present before he or she could arrive at an informed decision; he or she could not assume foreknowledge of the outcome or to know the best answer before having heard and considered the available alternatives. "People of intelligence," Isocrates maintained, "ought not to think that they have exact knowledge of what the result will be, but to be minded towards … contingencies as men [sic] who indeed exercise their best judgment, but are not sure what the future may hold in store" (1928-1945: 8-9). One way I try to show my students that they "ought not to think that they have exact knowledge" of an issue and that they ought to attempt to obtain a broader perspective before making their judgments is by insisting that they read and respond to differing viewpoints as they compose their texts and determine their thesis statements.

Isocrates also insisted that once a judgment had been reached in any particular situation or context it should not be considered resolved for good; he regarded all decisions as decisive and expedient, not static and final. In fact, for Isocrates, a well-functioning democracy depended on each citizen's willingness to allow his or her own opinions to be altered in the light of new evidence or new circumstances. "Even the gods," he said, made decisions by "debating among themselves" (1928-1945: 2). Like Isocrates, I encourage my students to take full advantage of the democratic freedoms to inquire and to debate, to change their minds and to try to change the minds of others. As he told one of his students, "Progress is made not through the agency of those that are satisfied with things as they are, but through those who correct, and have the courage constantly to change, anything which is not as it should be" (7). Such lessons about moral courage in the face of the apparent intractability of the status quo and a willingness to change one's mind in light of new data also seem to resonate well at an institution with a long history of graduates who have made courageous and often controversial decisions as leaders in industry, politics, and the military—General George C. Marshall being perhaps the most famous one. Like Isocrates, my pedagogical imperative is to create in my classes the conditions for ongoing dialogue and debate that are central to the functioning of a participatory democracy.

According to education professor Carl Glickman, contemporary American society is experiencing a "huge disconnect between education, public purpose, citizenship, and democracy" (2002: 7). Today, the average adult American "does not attend a single public, civic or community meeting a year," which has resulted in a "decline in citizen engagement in community, neighborhood or government affairs [of] more than 40 percent since the 1950s" (7). As Isocrates undoubtedly would have done, Glickman places the responsibility for this decline in civic participation squarely on the shoulders of educators. "America is on the cusp of a failed experiment," he says, "not a failure for corporations, individuals, or national wealth and not a failure of military or world power, but a failure to improve education of all citizens so that 'we the people' means just that, all of us" (7). In a political and social climate in which individuals do not often think that decisions of consequence are theirs to make and/or that public debate is nothing but a pointless shouting match among talking heads, I attempt to leverage the closely linked military and academic cultures at VMI to prepare my students to become citizens who are willing and able take part in the negotiation of the public good—a leverage that not all professors interested in teaching civic engagement have.

In his argument against "confusing democratic values with educational ones," Stanley Fish proclaims that while college professors "can reasonably set out to put [their] students in possession of a set of materials and equip them with a set of skills" they "have no chance at all … of determining what [the students'] behavior and values will be in those aspects of their lives that are not, in the strict sense of the word, academic" (2003: C5) by which he means bodies of knowledge and analytical skills. However, I concur with Joseph Gonzalez's Isocratean sentiment that transforming our students "is a natural consequence of doing our jobs well" (2008: A34). As Isocrates knew, and as Gonzalez reiterates, when students "become better scholars, both more knowledgeable and able to reject simplistic formulations, they naturally become better citizens … people ready to inquire, to think, and to engage with the world as they find it" (A34). Even Fish acknowledges that there may be a chance of merging educational goals with democratic values at institutions "where a certain homogeneity of culture is the advertised point of the enterprise and the students who enter that culture can be said to have self-selected" (2003: C5)—institutions such as VMI. What makes schools like VMI capable of pulling such contentions off, Fish claims, is that students who enroll at these colleges are not unwittingly compelled to endure lessons about democratic values in their classrooms. Instead, at institutions such as VMI, the students' experience "reinforces moral and civic choices they have made before matriculating" (C5).

Fish may be right that VMI can be so intentional in its educational approach to preparing cadets to become engaged citizens because we have the advantage of teaching a student body who has self-selected for just such an experience. Most VMI students do willingly choose to take part in a college environment that requires adherence to strict moral codes (VMI has a single-sanction policy for violations of its Honor Code, for example) and that overtly encourages civic engagement inside the classrooms and out. Having acknowledged these inclinations, more cadets are likely to be open to pedagogies that directly interrogate and openly advocate public service and citizenship—pedagogies such as my own. In fact, it was my awareness that most VMI cadets had some inkling that civic engagement involves more than voting every few years or using their purchasing power as consumers and had some desire to use their educations to do more than earn a nice paycheck that drew me to apply for a faculty position at VMI in the first place.

"Ready as Citizen-Soldiers"

When seeking students for his school, Isocrates openly proclaimed that he was looking for young men who had a genuine desire "to do some good in the world" (1928-1945: 269). To those who enrolled, he offered a course of study designed to provide them with an opportunity to "become better and worthier if they conceive[d] an ambition to speak well" (275). He taught his students rhetoric "for both public and private use—i.e., for those who sought to become politicians as well as for those who merely wanted to educate themselves and presumably to develop their capacity to assess the arguments made by politicians" (T. Poulakos 1997: 68). Consequently, he provided an education committed not so much to aiding his students' personal advancement or private commercial interests, but more importantly dedicated to those matters that concerned all society, such as advising leaders, promoting or opposing public actions (for example, declarations of war or peace), and establishing laws. He attempted to equip his students not only with the capacity for forming judgments and opinions but also with the ability to compose those opinions in such a way that they would have the chance to be considered and acted upon by other members of the polis. According to Isocrates, since his program primarily focused on the consumption and composition of "great and noble" discourses, he was able to "advance [his] pupils to a point where they [were] better men and where they [were] stronger in their thinking" (1928-1945: 185). As a result, "there was parsimony between Isocrates' theory and practice. He advocated an active role in the *polis* through which wisdom is put to the service of the common good, and that is what he and his students did their best to do" (Schiappa 1995: 55). It is this connection between theory and practice that I usually find myself able to achieve at VMI.

The official mission of the Virginia Military Institute is "to produce educated and honorable men and women, *prepared for the varied work of civil life*, imbued with love of learning, confident in the functions and attitudes of leadership, *possessing a high sense of public service*, advocates of the American Democracy and free enterprise system, and *ready as citizen-soldiers* to defend their country in time of national peril" (Virginia Military Institute 2009, emphasis added). Of course, as Fish laments, mission statements purporting to prepare students to be engaged citizens are not unique to military colleges such as VMI. On the contrary, as Bok explains, "preparing citizens still appears very bravely in most college catalogs" (2001). However, such preparation for citizenship "is very seldom talked about by faculties as a conscious aim, something to be implemented as a deliberate strategy for the development of their students." Yet VMI's claims about educating students to be active participants in the democratic process have more credibility than similar pronouncements at other institutions.

At VMI, a relatively large number of the faculty and administrators have prior military service, as do I. In addition, a military ethic is present in all the classrooms, as all tenure-track faculty members receive commissions in the Virginia Militia and therefore teach in uniform, and all the students come to class in uniform as well. At this intersection of academic and military cultures, professors feel comfortable implementing classroom strategies—such as my own Isocratean pedagogy—in a self-conscious effort to help prepare the students to become engaged citizens. And these approaches seem to be working. Like Isocrates, whose students included "the orators Isaeus, Lycurgus, and Hypereides; the historians Ephorus and Theopompus; the philosopher Speusippus; and the statesman and general Timotheus" (Norlin 1928: xxix), VMI boasts among its graduates numerous generals and Medal of Honor recipients, senators and congressmen, public speakers and college presidents. And while commissioning is no longer mandatory for all VMI cadets, just over 50 percent of recent graduates did choose to join the military, and many others demonstrated their "high sense of public service" by electing to become public servants in law enforcement, medicine, diplomacy, or education.

Of course I am not suggesting that VMI has cornered the market on distinguished alumni or that it is the only college educating community leaders. What I am suggesting is that the unique intersections between a military culture and an academic culture present at VMI provide an educational environment that is particularly conducive to counteracting the prevailing trend toward self-advancement and individual achievement in higher education and in our society in general. Isocrates attempted to temper this trend by directly connecting his students' lessons to their immediate engagement in a participatory democracy, which is why I believe that an Isocratean writing course is a good place to prepare the citizens we need to be responsible for and to advance our democracy. As I learned from Isocrates' precedent, "teachers ought to articulate the kinds of activities they want their students to perform outside the classroom, and they should design pedagogical techniques that develop skills in their students consistent with these future activities" (McComiskey 2000: 113). Fortunately for me, VMI provides an atmosphere in which I can openly articulate my belief that my students ought to use their education for moral and ethical aims, to negotiate the common good, and to defend our democracy, whether in a uniform or not.

Chapter 5
Rethinking the Culture Wars at the Naval Academy

Jeffrey S. Sychterz

The Military, Academe, and the Culture Wars

Americans generally do not think that academic and military cultures have much in common. Many on the right complain that the American higher educational system is dominated by radical leftists determined to indoctrinate students with an agenda that is, among other things, anti-war and anti-masculine. Many on the left consider the military not only a straightforward instrument of American imperialism, but also the foremost expression of an oppressive and violent patriarchal culture. Exacerbating this polarization is that the very people making these claims are often members of these opposing cultures, each regarding the other with suspicion, because they think of themselves, and their institutions, as committed to neutralizing the cultural influence and power of the other. Even in the middle space between these poles, the general population shares a similar, if not as extreme, point of view: the identity of the soldier and the academic are, in the modern professional world, antithetical; the one who fights the war and the one who analyzes its meaning, are, except in rare occasions, never the same. This view of polarized military and academic cultures plays a central role in an America increasingly divided into red states and blue states.

But, are academics and military service members really that divided? Does more divide them than unite them? Just as some political commentators have begun to question the very premise of a fundamentally divided America, I would suggest that a military/academic binary serves to mask deep similarities between the two cultures. Although military cadets and humanities graduate students may identify themselves as ideological opposites, many on both sides choose their future careers for remarkably similar reasons, which stem from an underlying system of shared values. I came to recognize and appreciate these similarities while teaching at the United States Naval Academy. The midshipmen I taught at the Academy seemed to share much more in common with my graduate school colleagues than with other college students I have taught, before or since.

While I was teaching English at the Naval Academy, a Marine Corps officer upstairs in the history department recommended I read Leo Braudy's *From Chivalry to Terrorism: War and the Changing Nature of Masculinity*. I found the book enlightening and enjoyable, but at the same time I was troubled by some of its sweeping generalities. One such generality particularly bothers me because it just doesn't fit my experience of military students. He writes, "Vietnam may have witnessed the last literary war … because in order to inspire widely accessible fiction or poetry, wars must be fought by men and women who are not professional soldiers. Certainly there will be memoirs and perhaps the occasional work of fiction, but compared to World Wars I and II, there will be very little literature" (Braudy 2003: 532). I find this reasoning problematic for a number of reasons. Although the initial claim may be true, Braudy sweepingly asserts that the identity of soldiers, or at least professional soldiers, "must be" inimical to the production of literature. Furthermore, Braudy makes his claim without much evidence; not that he couldn't have found plenty—a quick review of the most widely regarded works of war literature shows that almost of all of them were written by non-professional soldiers. And yet, what would that evidence really show? Most of these non-professional soldiers who produced war literature, such as Wilfred Owen and Ernest Hemingway, were not conscripted, but instead volunteered for military service. Volunteering to fight in a war blurs distinctions between professional and non-professional armies in a society that defines a professional force as one that is all-volunteer. Rather than suggesting that literature "must be" produced by non-professional soldiers, couldn't the conscription model of twentieth-century conflicts show instead that war "literature" requires a certain concentration of a population in uniform, say more than five million? It seems even more likely that the relatively limited scale of late twentieth-century conflicts is the primary cause of a tepid outpouring of war literature from conflicts in the Falkland Islands, Grenada, Kosovo, or the Persian Gulf. Perhaps these conflicts did not produce "literature" because they did not rise to the necessary level of existential threat that the World Wars or even the Vietnam War presented to the populations involved.

Even the precise historical moment that Braudy cites as the breakpoint between "literature" and "memoir" should point us to a more nuanced understanding of the relationship between literature and war. The Vietnam War era is generally acknowledged as the beginning of the ideological polarization of American society known as the culture wars. During that period a number of issues, most notably war, sexism and racism, came to a head in a dramatic fashion, supposedly dividing America into camps divided along traditionalist/conservative and progressive/liberal lines. Proponents of the culture war social model typically draw the boundary along moral and religious lines—between old-fashioned Christian values and a secular humanist worldview. And yet, the culture war's most iconic image is of flamboyant student protesters facing off against severe soldiers in Vietnam War protests. Frequently, the culture war is articulated as a standoff between antithetical academic and military cultures; and, especially during wartime, the university and the military base are used as convenient archetypes of its divided camps. During the current Iraq War, conservative rhetoric divides America between liberal elites, led by university professors, and "real Americans" serving their country in Iraq. In so many ways, such antagonism is genuine, if for no other reason than those invested in the culture wars have fervently drawn lines precisely along a liberal intellectual/conservative military partition. Such a partition means that idealistic young people looking to the future learn to define "service to country" according to their own, or their parent's, ideology. Generally, liberals serve their country in the classroom, conservatives on the battlefield. One group writes "literature" and the other fights wars. Therefore, to what extent does Braudy's claim of a fundamental split between the writers of literature and the fighters of wars rest upon this recent history, as well as upon the very claims of the culture wars? Such a split might be real, but does it reflect a recent sociological construction rather than a more enduring fissure between institutional cultures? In other words, are professional warriors fundamentally ill-suited for writing, or even teaching, literature? Can we really not expect engaging and introspective literature to emerge from military culture?

Despite this line of questioning, I probably would have accepted Braudy's claim not so long ago, even though my entire life has been spent crisscrossing military and academic institutions. Most of my childhood was spent as a Navy brat in military schools, except for a brief, but formative, experience in middle and early high school, when my family lived in a large university town while my father went back to college. For my own college experience, I went to the United States Naval Academy. After graduating in 1991, I served eight years as a Naval Flight Officer flying Maritime Patrol missions. In 1999, I left active duty to pursue a PhD in English literature at the University of Illinois at Urbana-Champaign; currently I am an Assistant Professor of English at Fayetteville State University in North Carolina. Such a history might suggest a comfortable melding of academic and military cultures in one identity, especially considering that while at the Academy I majored in English, and while in graduate school I served in the naval reserves. However, throughout my traversing of these institutional cultures, I too shared a strong belief in their incommensurability. Like many Americans, I felt that academia and the military were conflicting spaces, separated by incompatible world views. One does not meet many Democrats, let alone confessed liberals, in the military; and conversely, one does not meet many Republicans in a university English department. Certainly, there are a few exceptions in both cases, but their appearance is rare enough to confirm that the realms of liberal education and military service are indeed ideologically opposed. In graduate school, putting on my uniform to go to drill weekend felt strongly like climbing into a different skin and becoming someone else, even though I had worn a uniform daily for the previous twelve years. I felt that I could move between identities, but could never inhabit them at the same time; they were just too different. Different, that is, until I had the opportunity to return to the Naval Academy where I simultaneously inhabited both subject positions, the uniformed and the gowned. It was there that the midshipmen taught me that what had first drawn me to the military was precisely the same sense of service that drew me to academia.

Although I returned to the English classroom that I had prepared for in graduate school, I returned in a military capacity: as a recalled reservist on three year military orders. Not only did I teach midshipmen at a military academy, I did so while wearing the uniform of a naval officer. My students called me sir, and I was not Doctor Sychterz, but instead Lieutenant Commander Sychterz. At first, the new job was difficult to wrap my mind around. With the uniform on, I felt out of place; I should have been going to an aircraft hanger for a departmental meeting, not to a classroom to teach literature. I struggled for the right mindset and tone, and the uniforms "ranged in columns before me" only heightened the disorientation. I should not have felt that way. After all, I had just spent the last six years in the classroom, both as a student and as an instructor, and now felt quite comfortable in it. Furthermore, I had returned to my alma mater, where, as a midshipman, I had read and discussed Shakespeare and the Beat poets while in uniform. The problem seemed to be that in uniform I didn't think like a teacher but like a naval officer. The uniform not only symbolized a different mode of thought, it almost required it. The classroom is a place where one can contemplate complex problems without the need to act on them. It encourages students to take up tough social, ethical, political and philosophical problems, not to solve them but to complicate them with an increasing series of questions and points of view. Teaching, for me, is in part helping students develop tools to discover the complexity within seemingly simple issues. The uniform, on the other hand, symbolized rigidity of thought and the reduction of a complex issue to a simple solution: getting a weapon on target. The classroom encourages the questioning of authority, the uniform quashes it.

Over time, however, that hard-line distinction between identities faded, mostly through the inevitable familiarity bred by repeatedly teaching in uniform, but partly because the midshipmen I encountered did not conform to the stereotype I had developed around the uniform. They were very energetic and eager to talk; they relished classroom discussions. By comparison, the students at the University of Illinois, where I taught as a graduate teaching assistant, were taciturn and uninterested in the direction of the course. I had many classes at Illinois that were very quiet, because the students had nothing to say. They had done the reading—quizzes and tests confirmed that—they just felt indifferent toward and untouched by what they read. Meanwhile, in three years of teaching at the Academy, I never once had a quiet class of midshipmen. Even when they have not done the reading, midshipmen have something to say. And what they have to say is not simply official wisdom passed down through the chain of command, but instead critical and thought provoking ideas. Midshipmen proved themselves to be more supple critical thinkers than students I had taught in Illinois. Despite what I had expected, midshipmen were more willing to question received opinion and seek out complex situations and moral ambiguities.

"Bartleby" and Midshipmen

I was first made aware of this greater receptivity for moral complexity in teaching Herman Melville's "Bartleby the Scrivener." Melville's short story tells of a lawyer bedeviled by a singularly odd copyist at a law firm who "would prefer not to" do any work. Despite the lawyer's attempts to understand and even save Bartleby, the copyist sinks increasingly into a morbid torpor until he gives up on life and dies. This story has come to symbolize for me so much of my entire tour at the Naval Academy, not because Bartleby shares any characteristics with the midshipmen—although the crypt-like office setting that entombs Bartleby can seem awfully familiar to a plebe (a freshman) living through what are known at the Academy as the Dark Ages of winter. Instead, this story will always remind me of the Academy because I had my first real success teaching it there, and that success had everything to do with the values to which a midshipman aspires. I love this story, and teach it whenever I can; however, teaching it at the University of Illinois was very frustrating. It is, I must admit, a difficult story for students because we know almost nothing about Bartleby, and what we do know is filtered through the lawyer/narrator, whose own vanity and misanthropy obscure any real understanding of the unfortunate scrivener. That lack of understanding actually angers some students—they hate Bartleby because they cannot understand why he does what he does. They solve that anger by simplifying the story's lesson: Bartleby is a bad employee; unproductive employees hurt a business's bottom line; the lawyer should fire Bartleby. That the lawyer cannot bring himself to fire Bartleby demonstrates only the lawyer's cowardice and incompetence. The students at Illinois felt that anger most acutely. They strongly identified with the lawyer's plight and condemned Bartleby most harshly. They held firm to their simple summation of the story even as we went through its details, including some of its evidence in favor of Bartleby, such as the story's Christian themes, the soul-destroying setting and nature of Bartleby's work, and even the lawyer's own self-interested narrow-mindedness. Despite my best efforts, most students could not conceive of an alternate frame through which to consider the story's situation. They tended to see even the lawyer's own abortive attempts to understand Bartleby through a religious or humanistic context as signs of weakness. When I asked if the lawyer had any responsibility to Bartleby as a human being, not just as an employee, the answer was overwhelmingly no; Bartleby had to take responsibility for his own actions. For most, it was a workplace story that called for an MBA solution. The lawyer's concern for Bartleby should extend no further than the quality of his work. Any other consideration only compromises workplace efficiency.

Initially, Naval Academy midshipmen approached Melville's story in a similar manner: the enigmatic scrivener frustrated and confused them, and a few personally disliked him. They too argued that Bartleby is ultimately answerable for his own actions, and that the lawyer has no responsibility for him once he refuses all work. Yet, the midshipman's own chosen career path offers a workplace model very different from the typical boss-employee one through which the Illinois students made sense of the story. So, when midshipmen also suggested that firing Bartleby was the only solution to the story's problem, I asked if they would feel the same way if Bartleby was not a scrivener, but instead Petty Officer Bartleby and the lawyer was a lieutenant in the US Navy. Suddenly their answers changed; in that case,

Lieutenant Lawyer would have to counsel P.O. Bartleby and try to ascertain the causes of his bizarre behavior. Yes, if Bartleby continued in his passive resistance he would have to be separated, but, in the meantime, the officer is responsible for his subordinate's actions and must take steps to help him—even if Bartleby is suffering from a personal problem outside of work, and even if Bartleby resists attempts to reach him. This shifting of the problem from one of management to one of leadership opened up the story and the midshipmen began to critique the lawyer's failings as a leader. According to the leadership model, an officer has a responsibility toward his people that far exceeds the bounds of the workplace: a good officer truly cares for his people even when they screw up. Now, one might point out that such care is still done in the ultimate service of the mission, as some midshipmen certainly did, but as anyone familiar with military service can attest to, the caring is still real. It is the ultimate paradox of combat leadership: the leader is asked to care deeply for the troops that he someday might have to lead to their deaths. That love has a long history and is palpable in war literature from *The Iliad* to Siegfried Sassoon's poetry. It helped midshipmen see some of the larger human questions raised by Melville's work, and it changed their perspective.

I tend to follow this story with a paper assignment. I tell the students that the lawyer stands accused of Bartleby's death in "a court of humanity," (the lawyer contributed to Bartleby's death through his own lack of care for the scrivener). The students are to pretend they are lawyers arguing the case. They can choose to argue on behalf of either the defense or the prosecution. The assignment revealed stark differences between the civilian and military institutions. Most of the civilian students at Illinois chose to defend the lawyer, arguing personal responsibility, while midshipmen overwhelmingly assumed the role of prosecutor and sought to prove the lawyer's guilt through his lack of concern for other human beings.

Now, one might argue that these different responses prove very little, because the midshipmen were given another option through which to bring meaning to the story. The leadership model was generally not available to students at Illinois, and if it had been available they would have approached the story in a similar way. That is certainly true, but it matters that the model was available at one institution and not the other. The very existence of the leadership model is what draws many students to the academy in the first place, and even if it did not, leadership plays such a pivotal role in a midshipman's education that it quickly provides the frame through which she approaches the world. So, the very fact that the institution makes such a frame available does point to a distinct difference between the belief structures of the civilian students and their military counterparts. In fact, I did try to push the Illinois students to think about the lawyer-Bartleby relationship in contexts other than employer-employee, but the economic one was primary and pretty much insurmountable. To them, Bartleby is just an employee, a commodity to be utilized like the copy machine that has taken his place. Their goal after college is to get a good job making good money; Bartleby only matters in whether he promotes or hinders their acquiring of capital. For the midshipmen, on the other hand, the military model quickly trumped the economic one and allowed them to consider Bartleby as a human being. From their perspective, they have subordinated their personal goals to those of the nation, swearing to protect people like Bartleby. They are motivated by honor, ethics, and humanity much more than material gain. From this standpoint, Bartleby is not an obstacle, he is the mission.

"Don't Ask, Don't Tell"

Generally, the responses in the Bartleby paper contradict broad institutional stereotypes that the American public holds of the liberal college student and the conservative cadet, but they accurately reflect a greater trend in my classroom experiences. Midshipmen at the Naval Academy tended to be somewhat more liberal on social issues than students I taught at the University of Illinois or even Fayetteville State University. While occasionally I had students at Illinois who conformed to the radical liberal stereotype, like the outspoken PETA advocate with the impressively tall Mohawk, many more students were conservative Christians, such as the business student who dyed his hair snow white and was a member of the Campus Crusade for Christ and the Fellowship of Christian Athletes. Certainly, Illinois has its share of student protests, but one of the most prominent is the annual pro-life vigil. Such religious conservatism is even more prominent at FSU, because it is a southern Historically Black College. There, students routinely use Christian doctrine to back up arguments in the classroom. I was quite surprised the first time students said "Amen" after I had quoted from the Bible. After all, I was only making clear a literary allusion in a poem. Because of this greater religious commitment, students at the two civilian institutions tended to take more conservative stands on cultural issues than the midshipmen at the Naval Academy.[1] Even on a defining issue like homosexuality in the military, the midshipmen were surprisingly open-minded.

On October 21, 2005, the Equality Ride came to the Naval Academy to protest the military's "don't ask, don't tell" policy. The Equality Ride is an annual bus trip to protest colleges and universities that discriminate against gay, lesbian, bisexual and transgender people. Although they have incited drama at some schools,

1 Granted, these are my personal observations of the classroom environment, and not backed by statistics; however, my colleagues at all three institutions have shared similar impressions of ideological trends among the students. I must also point out that Illinois and Fayetteville are by no means representative of the entire US higher educational system. Their greater social conservatism does reflect demographics; one is a Midwestern state school and the other a southern Historically Black College. Even the military academies are affected by those demographics. The Air Force Academy tends to draw more students from the Midwest, and therefore more conservative Christians—as the 2005 evangelical scandal demonstrates. Both West Point and Annapolis tend to attract more students from the coasts, which tend to be more secular and liberal, but West Point seems to draw more from the South, so I expect that the students at the Naval Academy may be slightly more liberal than either their Army or Air Force associates.

including arrests at West Point in 2006, the 2005 protest at the Naval Academy proceeded peacefully, and consisted of nothing more than several people taking the time to talk to midshipmen about the policy. As I looked out my classroom window that day I could see several protestors dispersed across the campus—I could easily identify their colorful "Equality Ride" t-shirts through the drizzle because they offered a striking contrast to the black uniforms moving en masse from lunch to class. My students could see them too, and they asked if we could invite one of the protestors into our class to explain their opposition to the Department of Defense policy. I said, no, assuming the request came from a desire to badger the protestors and best them in an argument. But the students insisted, explaining that they wanted to hear and consider different sides of the issue, and I soon realized their request was genuine and polite. After all, the classroom provided a safe and somewhat anonymous space where the students could talk to the protestors without attracting negative attention to themselves. Nevertheless, I still declined because I was unsure of the ramifications; after all, I was on active duty at the time. However, I told them that we could take a break from the assigned reading and discuss the "don't ask, don't tell" policy ourselves. That class discussion was one of the most enlightening moments of my three years teaching at the Academy, and it overturned many assumptions I had about midshipmen.

Of the students who spoke, only two supported the policy. The rest opposed the ban on homosexuals serving openly in the military. If I was surprised by the general sway of opinion, the midshipmen were even more surprised. Each one who opposed the policy had believed that he was the only one at the Academy who felt that way. This situation was best summed up by one plebe, who recounted a discussion he had overheard at lunch. One firstie (senior) asked another what he thought of the Equality Ride protest. The second said, "I'd rather hear what you think, because my opinion is different from most people's." The first then suggested that he thought the ban needed to go, to which the second responded, "Wow, I thought I was the only one." Midshipmen knew what opinion they were supposed to hold, and therefore thought that military service members naturally conformed their opinions to the official party line. Those who thought differently were probably afraid to express their opinions because they thought such dissent would be frowned upon as an unacceptable aberration. The discussion we had in the classroom reflected a dawning awareness by all of us in the room that Academy midshipmen do not hold monolithic views on homosexuality and the military.

Just as instructive were the reasons used to support both sides of the debate. The two outspoken supporters of the ban relied primarily on biblical arguments. They contended that homosexuality was immoral because it perverted natural law—that is, a nature formed by God's law—and was specifically prohibited by God. They felt that the government must not in any way condone homosexuality because it would lead to the breakdown of civil society. Perhaps more important than what they did argue, was what they did not; they did not use the "unit cohesion" and "good order and discipline" arguments so central in forming the policy in 1993. Instead, it was the students who felt the ban was outdated who cited military effectiveness. They contended that the military already bars its members from fraternization and sexual harassment, and these regulations are independent of gender or sexual orientation; even heterosexual members are prohibited from using their sexuality to create a hostile or offensive workplace. Any unwanted sexual advances are prejudicial to good order and discipline. According to one midshipman who emailed in support of the Equality Ride,

> There are a number of midshipmen, including myself, who support serving with any citizen who agrees to abide by the set standards of conduct regarding fraternization within units. This was not a new standard introduced by Clinton in the 1990s; this was introduced when women began serving decades ago. It is meant to maintain good order and discipline within the services. These rules entail that officers and enlisted may not date, nor have unduly familiar relationships. These rules also apply to members of the same command, or unit. (qtd. in Reitan 2005)

The students in my classroom reached the same consensus. They felt that the "don't ask, don't tell" policy was unnecessary. According to this argument, the military should be a sex-free zone: all military members should be barred from talking about their sexuality in the workplace. They effectively made a non-issue of the distinction between an "open" or "closeted" sexual identity. They interpreted fraternization and sexual harassment statutes to mean that even heterosexuals should closet their sexuality in the workplace. Furthermore, they felt a service member had the right to be open about his or her sexuality outside of the workplace. After all, one midshipman added, what happens in the bedroom has no place in, and therefore no impact on, the workplace, period. Not a single midshipman argued for gay or equal rights. If they saw the policy as prejudicial, they saw it as prejudicial to military effectiveness, because it barred otherwise competent soldiers and sailors from serving in the military based solely on their sexual orientation, an aspect of identity they felt didn't belong in the workplace in the first place.

This attitude is much more accepting of homosexuality than what I encountered from students at either the University of Illinois or Fayetteville State University. While I have had a couple of openly homosexual students at FSU, many more are quite vocal in their opposition to homosexuality. One memorable incident occurred during a student's persuasive speech opposing gay marriage. He began his presentation by circulating some pictures—he handed them out on the opposite side of the room from where I was sitting. I became nervous as just about every student recoiled in disgust when they looked at the pictures, some saying, "that's disgusting" and "gross." When the pictures finally reached me I was relieved to find that they only depicted men holding hands and kissing; the students had reacted as if they were X rated. Six years ago as a graduate TA, I encountered similar uneasiness. The Illinois students I had were also very uncomfortable with homosexual themes and representations in literature, and they struggled to understand opinions that were different from their own. One work I have taught that deals with the complexity of gender and sexual identity is *Regeneration*, Pat Barker's fictionalized account of the World War I hospital where Siegfried Sassoon and Wilfred Owen were treated for shell shock. In it, Sassoon's doctor, W.H.R. Rivers, theorizes that rigid codes of British masculinity, and even heterosexuality, play a contributing factor in the soldiers' psychological breakdowns. Sassoon was sent to the hospital for political reasons, and Rivers wonders if Sassoon's homosexuality helped him avoid a breakdown because rather than repressing his trauma his more flexible emotional identity allowed him to emotionally confront his wartime memories. After completing the novel and discussing these scenes in an Introduction to Fiction course at Illinois, every student who wrote about the issue claimed that the novel argues that men were breaking down in the trenches because they were homosexual and, therefore, emotionally weak. In contrast, when I taught Barker's novel at the Academy, while a couple of midshipmen were dismayed by the romantic tension they identified between Owen and Sassoon, most were nonplussed and discussed it as they would any other literary topic.

Service and Idealism

Homosexuality was not the only issue on which I found the midshipman to be more open and critically minded. When it comes to issues of the nation, the military, and national security, midshipmen are indeed a patriotic bunch, but even on those topics they are diverse and thoughtful. One of the classes I enjoyed most while teaching at the Naval Academy was the Introduction to War Literature. I was initially apprehensive about it, even though I had written my dissertation on war poetry. I expected the class would be filled with students who wanted to read rousing war narratives. I was worried that they expected Clive Cussler and Tom Clancy, while I would give them Wilfred Owen and Kurt Vonnegut. My worries turned out to be unfounded, because while the students were almost completely uninterested in questions of form or genre, they dove right into the ethical complexities of warfare and the warrior identity. We had detailed and substantive discussions on issues such as honor, masculinity, war crimes and even post-traumatic stress disorder. Furthermore, they distrusted overly-romanticized representations of warfare, even though none of them had any combat experience. They unanimously disparaged *Saving Private Ryan* as too sentimental, even criticizing the tear-jerking scene when the older Ryan asks his wife if he has led a good life, one that sufficiently compensates for the deaths that made it possible. They complained that death in war is brutal and random, and Spielberg tried to erase that fact by wrapping Captain Miller's death with an aura of heroism. A couple of students were quite ready to go further and at least entertain the possibility that no moral basis existed for warfare at all. One student, Jim, suggested that gang warfare is more moral than a war between nations, because gangsters are openly ruthless while governments always hide their true intentions. "After all," he said, "when the godfather wants someone eliminated, he says 'kill that guy, because he's muscling in on my business,' but a nation always lies and has to use patriotism and nationalism to get people to think they're protecting the country." Despite his cynical view of the nation state, as well as some worries about post-traumatic stress disorder, he elected to join the Marine Corps because he wanted to serve on the front lines in the infantry. I asked him how he could justify his service selection in light of his feelings about war and the nation, and he shrugged his shoulders. He felt it was something he had to do.

In fact, I was always interested in what midshipmen wanted to do after graduation, and why they had come to the Academy in the first place. I received almost as many answers as there are midshipmen, but their motivations almost always fell under the umbrella of one overriding concern: they wanted their lives to have meaning. Most wanted to make a difference in the world, and they saw the military as the best, or perhaps the only, option to have that impact. Many young people join the military because they believe deeply in human rights and see the military as the best way to protect those rights. In fact, one plebe wondered if he made the right choice coming to the Academy rather than enlisting because he wanted to get over to Iraq as quickly as possible to help "those poor Iraqi people." Jim, the war literature Marine, probably would cite this as evidence of crass manipulation by the government, but the plebe's idealism and altruism were genuine. I would be willing to bet that Jim had once felt the same way and his idealistic view of the military has remained strong, even as his faith in nationalism has waned. The military would still be, in his eyes, an avenue for the strong to protect the weak, even if he thought the US government manipulated those altruistic motives for its own economic interest. As a combat officer on the front lines, he could mitigate the damage through ethical and moral leadership. Many midshipmen feel they could make more of a difference on what they call "the pointy end of the spear." Behind the lines they are just one more citizen whose voice has to compete with 300 million other voices. As leaders of a small team in combat they can make decisions that directly affect people's lives, and perhaps help to shape US force projection in a more productive and ethical way.

Overall, the idealism is what struck me most about midshipmen in my three years teaching at the Naval Academy. Our literature discussions tended to be more fulfilling than elsewhere I have taught because midshipmen wanted to engage with explicitly ethical questions, and an appeal to leadership could always get them to think about the human questions at issue in a text. One of my plebes responded to a Vietnam War poem (W.D. Ehrhart's *Guerrilla War*) by updating its situation to his possible future: "What do you do if you're guarding a checkpoint in Iraq and a pregnant woman comes running at you and ignores commands to stop?" Nine out of ten of his classmates said that you don't shoot her; in all likelihood she is running to you for help. Besides, it would be better to die than to risk killing a civilian. Dying is part of the job, they argued; the soldier or sailor goes in harm's way to ensure the civilian does not have to.

Midshipmen see this willingness to sacrifice their lives to protect another's as what differentiates themselves from other college students. They have chosen a career of sacrifice and service over one of profit. Students at other institutions increasingly come to college because a college degree means a larger income; this pecuniary motive is intertwined with the transformation of the university from a institution that creates enlightened thinkers and citizens to one that produces employees for the corporate workforce. This gradual corporatization is not only shrinking the traditional liberal education offerings, it is also changing the tenor of those classes that remain. Take the earlier example of "Bartleby the Scrivener": Illinois students could not think beyond an economic rationale for the lawyer's treatment of Bartleby because the economic rationale dominates their own life calculations. They see themselves as starting in middle management and moving up from there. Their goals include being millionaires by age 30 and CEOs of Fortune 500 companies soon after that. That's what students frequently told me when I taught there. They, therefore, immediately see themselves as the lawyer, not the scrivener. Bartleby, as an employee, is just another tool of wealth production, just like the college degree itself. If the tool isn't performing its task, you get rid of it and find another one.

It is this type of thinking that angers many midshipmen. In fact, more midshipmen cited the money-obsessed college student, and not the stereotypical liberal protestor, as a reason for coming to the Naval Academy rather than another school. While other college students sell out, midshipmen stay true to a higher calling. President Obama identified these same attributes when speaking at the Naval Academy's 2009 commencement. He praised the graduating midshipmen, because,

> In a culture where so many chase the outward markers of success that can so often lead us astray—the titles and status, the materialism and money, the fame and popularity—these Americans have embraced the virtues that we need most right now: self-discipline over self-interest; work over comfort; and character over celebrity.
>
> After an era when so many institutions and individuals acted with such greed and recklessness, it's no wonder that our military remains the most trusted institution in our nation. (Obama 2009)

President Obama's praise contains a subtext: virtues such as selflessness have become increasingly rare in the US. Our culture is dominated by greed and selfishness, and the military not only offers an escape from this national trend, it can also provide its cure. That kind of self-abnegating idealism appeals to most midshipmen.

This kind of motivation sounds suspiciously like what draws many academics to the classroom as well: "We cultivated the sustaining fiction that it was a high-minded devotion to learning that had brought us to graduate school, while so many of our peers had sold out by studying law or business, but with high-mindedness came the pose of martyrdom. We surrendered rather easily to yet another romantic notion: that meaning is to be found only in misery" (Samet 2007: 35-6).

Midshipmen buy into this fiction as well, and take secret comfort in the hardships of life at a military academy, some petty, some not. Like a graduate student trying to write a dissertation, the midshipman feels validated by suffering; it means she has not sold out.

The quote comes from Elizabeth Samet's own meditations on teaching English at the Military Academy at West Point. Yet, even scholars not associated with military academies have noticed parallels in the motivations of what seem to be ideologically disparate cultures. At an academic conference in 2006, I happened to fall into conversation with two self-confessed Marxists. They asked me what it was like teaching at the Naval Academy and I told them I found it much easier and more enjoyable than at the University of Illinois. I then shared my observations about midshipmen's idealism. Rather than expressing surprise, they said this confirmed a suspicion that they had. They had been wondering where all the young Marxists and socialists were today, when one said, "They are all over in Iraq." By this statement, he meant, I think, that the military seems to offer the only escape from corporate America for a young idealist, particularly one who is socioeconomically disadvantaged. US society today lacks a large scale, socially legitimized leftist movement: labor unions have been losing ground due to the loss of an industrial base, and teaching requires an advanced degree. Besides, higher education has been tarnished by accusations of elitism and lacks the same cultural authority as the military. Therefore, volunteering to fight in Iraq is not just an individual response to the terrorist attacks of 2001, but also an individual solution to the materialism and vapidity of a worn-out consumer culture.

The suggestion of socialists in the military is not as crazy as it may sound. I had one student, Dana, who wanted a career in the Navy because he found it to be the most truly socialist institution in America. In some ways, a military academy is a more social and socialistic institution than the average college. While most midshipmen pine for the extracurricular excitement available at a civilian college, they fail to notice that their own social lives are much richer. A student at a big university can feel lost, isolated and alone, but at the Academy, midshipmen all work together on a shared goal, a goal that is in essence their own personal development. Ironically, therefore, the inner workings of the Brigade of Midshipmen more closely resembles socialism, while the large state university system more closely represents the individualism and commercialism of capitalism. Midshipmen work together for a common goal that melds one's individual identity into that of the community. Communal progress and individual development go hand-in-hand. Meanwhile, at the state university, the college student is focused almost entirely on himself. His goals are personal and have little connection to the educational institution or to the educational experience of his peers. Many times in conferences at Illinois, I could feel students calculating every moment of their college experiences, wanting to know how much reciting this poem would add to their future wages. The Academy offers many midshipmen a way out of this narcissistic individualism and materialism.

It is precisely through this opposition to American materialism that military conservatism and academic liberalism should be able to find some common ground. Both institutions, in my experience, draw idealists who find themselves alienated from a self-centered culture that commodifies human relationships and cultural values. For the academic professional, teaching and publishing directly encounters ideas in the hope of changing them for the better; for the military professional, submitting selflessly to the common good, and even defending the culture from which one feels alienated, can transform that culture for the better. Interestingly, both sides tend to identify this broad middle ground of selfish wealth accruement with the other's ideology. Conservatives blame liberals for denigrating higher values of honor, virtue and sacrifice in the name of personal self-interest, while liberals blame conservatives for reducing human beings to a commodity that can be bought and sold in the name of greater wealth production. Yet, the two institutions often identified as the vanguards of these political extremes share a common alienation that cuts across ideological boundaries. Quite a number of young people who pursue academic or military careers do so as a way of opting out of the material culture of Corporate America. They want their lives to mean something more, and to be measured in units more transcendent than dollars. They may define service differently, one serving the nation and the other humanity, but they still choose service over wealth.

I do not mean to suggest that the Naval Academy is a stronghold of enlightened liberal thinkers. It is still a deeply conservative institution, where students are committed to molding themselves after a patriarchal masculine identity, and remain suspicious of anything feminine. Even many female students are quick to distance themselves from "femi-Nazis," and a few try to be more masculine than the males. Furthermore, the military has tended to lag behind the greater American society on the integration of women and homosexuals. Many would argue that the military took the lead on racial equality because of Executive Order 9981, which desegregated the armed forces in 1948, six years before Brown versus the Board of Education. But even racial equality took much longer to achieve than we tend to remember: "The desegregation of the armed forces followed a course that roughly paralleled the vicissitudes of the civil-rights movement in national politics" (Mershon and Schlossman 1998: 309). Even after official desegregation, "the racial views that service personnel brought with them from civilian society often remained unaffected by anything that happened with the armed forces" (320). In fact, the US military continued to suffer greatly from racial conflict up through the Vietnam War.

Furthermore, I did not encounter the same sense of idealism and commitment to human rights when I was a midshipman at the Academy in the late 1980s and early 1990s. Then, the stereotypical culture wars alignment seemed firmly entrenched. The idealism may have been there and I just missed it; however, I think what I witnessed at the Naval Academy between 2005 and 2008 may represent the beginning of a sea change in military culture. As we move forward in the twenty-first century, militaries around the world are becoming more associated with protecting human and civil rights. In this respect, the United States is lagging behind, but rapidly catching up. The officers I met from Britain, Canada and Australia tended to define their role as humanitarians and peacekeepers first, and protectors of the nation second. Many on the left may still consider the military in the US as primarily a tool of American hegemony and imperialism, but, judging from the attitudes of tomorrow's naval officers, that may be changing. I even expect that the next great work of war literature might just come from one of these future officers now studying at a military academy.

Chapter 6

Literature, Identity, and Officership

Andrea Trocha-Van Nort

> What's Hecuba to him, or he to Hecuba,
> That he should weep for her? What would he do,
> Had he the motive and the cue for passion
> That I have? He would drown the stage with tears
> And cleave the general ear with horrid speech,
> Make mad the guilty and appall the free,
> Confound the ignorant, and amaze indeed
> The very faculties of eyes and ears. Yet I,
> A dull and muddy-mettled rascal, peak,
> Like John-a-dreams, unpregnant of my cause,
> And can say nothing; no, not for a king,
> Upon whose property and most dear life
> A damn'd defeat was made. Am I a coward?

Hamlet's Act II, Scene 2 speech on cowardice and motive raises questions about our unwillingness to act, especially when uncertain of the soundness of a cause (Shakespeare 2006). A cadet's loss of identity upon arriving at the US Air Force Academy (USAFA), having his or her worldview drawn into question, and the ongoing shaping of a commitment to the Academy require faculty to assist cadets in restructuring that identity. Time spent in military training and in academic pursuits must include opportunities for cadets to reflect upon their prior views of the military, the window upon duty that the Academy offers, and the intersections between the two, upon which will be cemented the foundations of the future officer's commitment to the armed forces. Key in this growth, the different domains of the humanities allow cadets opportunities to discover themselves. As our service academies (the USAFA, US Military Academy, US Naval Academy, and the US Coast Guard Academy) unavoidably push toward greater specialization in technical fields, decision-makers may begin to reserve a less active role to English and the Humanities among core courses. Vigilance in preserving a vital core liberal arts program supports the service academies' officer development programs as one of their most precious assets, building character as well as bridges of understanding. Academic dynamism can be maintained in the Humanities by recognizing the influence a carefully wrought program may have with regard to officer development as well as to the students' expectations of their own active duty career. After a brief description of my own passage into the service academy world, we will explore service academy criteria for selecting texts for study, preparations geared toward officer development, and approaches to make texts and coursework meaningful to cadets' recasting of their identity. Embedded in this study are examples in which the presence of a civilian may be advantageous in the service academy classroom.

From France to Colorado Springs

Though unofficially, I felt over my fifteen-year career in France teaching English, management, and economics that I was nonetheless an ambassador of America, its culture as well as its politics. At first I was highly offended when my svelte French students would offer scathing opinions on obesity in America, or on gun laws, the death penalty, and many more of our other "crimes." Sensing a personal attack, I at first returned their fire with my own polemic on the various hypocrisies of French culture and law, such as growing rates of obesity in the *banlieus* (government housing projects in the suburbs), the rising violent crime committed via knives, bats, and so forth, and even the ambiguity over the process by which the death penalty was erased from the Fifth Republic's books. These retorts rarely met with anything but increased anti-Americanism and a sense of "me against them." I quickly learned that if the students were going to trust me, they had to believe that I was broadminded enough to see things from their perspective, and that I cared about (not simply tolerated) their culture. In other words, pride had to give way to dialogue, and my own positions on the issues that interested them could only come forward in the form of a question. I realized as well that my students were identifying what they appreciated about their European heritage by simultaneously exploring and demonizing the more obvious facets of the United States': rapacious corporations as opposed to their stakeholder capitalism and culture of worker entitlement; unconsidered materialism versus more realistic consumerism; a disregard for all things ecological compared to the care taken in Europe with packaging, use of fossil fuels, and in household sorting of refuse; and finally, our unilateral decision-making as opposed to the European aim for consensus amongst its members—in short, everything that classifies us, for them, as *les grands enfants*, or over-grown children, which they have called us since World War II. Once I had understood their needs, indeed academic ones, I became a more effective unofficial ambassador, a result I could see in the new approaches students began to take—comparing, seeking similarities, hoping to find ideas they could apply to their own lives, instead of simply dismissing the "empire of disdain."

Post-9/11, I perceived various incidents as signs that I could be more useful to my country in a more official capacity than in my then current job as a *maître de conférences*, or associate professor, in a French university. I had survived the challenge of teaching American civilization courses to auditoriums full of first-year overwhelmingly anti-American students, had succeeded in seeing my graduate students well-prepared to take on challenging internships in the US, and had even published a student manual in which I was able to share and elaborate on those better ideas coming from the US economic incubator. I felt a sense of accomplishment in each of these undertakings. Still, anti-Americanism as manifested through my husband's *banlieu* fourth-graders, when they cheered in the classroom at the fall of the twin towers, the disappointing loss of Franco-American *entente*, and my research director's sudden tears when describing his memories of the Americans' liberation of Normandy, all made me wonder if I couldn't be more useful to my country in rebuilding that *entente*, with France and other traditional allies, from home. Couldn't what I had learned in France and Europe be of use to my countrymen and women? Especially my country's youth? The service academy came to mind, and, though I wasn't certain that my teaching or the disciplinary knowledge that I had acquired would pertain, I felt certain that the rigor I had learned from the French system and the reflexes I had developed would allow me to be an effective service academy teacher. For the purposes of this chapter, I can say that the central pedagogical lesson that my French teaching experience offered is recognizing that one's lesson plan and even course aims are meaningless bits of paperwork if they do not take into account the student body's hopes, needs, and perspectives. The ability to walk into Plato's cave with each class, to read behavior attentively, to ask the right questions with an earnest desire for spontaneous and thought-provoking answers: this is what I developed abroad and have tried to employ here, in my more official position at the USAFA.

A Collaborative Curriculum

What I noticed immediately upon arriving is that the institutional mission of the service academy necessarily focuses course development. Whereas in a civilian institution greater freedom exists regarding the direction of class discussion topics, a military environment's schedules (Monday through Saturday at USAFA, a minimum of 18 credit hours per semester, and mandatory intercollegiate sports or intramurals every afternoon), and limits placed on students' attention spans (responsibilities within the squadrons, training of freshmen, the push to volunteer for civic duties) force a substantially condensed yet intellectually compelling study of themes and texts over a semester. This is clearly different from French upper-education courses, where what is discussed in class rarely finds its way into a test question; the material discussed is to spur students toward intellectual inquiry, outside work for which they will be held accountable on exams. Instead, a military academy instructor weighs the texts to be studied in class carefully, as the discussions that evolve, combined with direction, questions, and depth provided by the teacher, are the sources from which one develops assessments. First and foremost, the content needs to be relevant to the institution's core values as well as to what the students see as realistic choices with which they will be confronted in the future. One must question the ends of studying a given text, reflect upon discussions of the particular issues presented, its sources, and its agents, as well as draw together one's own vision of the work, an interpretation that will be valuable to the students in their future careers.

Instructors conduct this advance work in groups for the core texts, something very new to me, as we always worked alone in France. As one might imagine, our group meetings are opportunities to talk about our common interests, language and literature, and to establish stronger bonds between our colleagues, military and civilian. In addition, though course prep is usually carried out on one's own for texts supporting the English major and enrichment courses, a teacher proposing a course must collaborate with others regarding the needs of the department and larger institution. The group's input enables the instructor to develop several approaches to explore in class. We must also determine which angle will offer the students the greatest possibilities for discussions, as class time is short (we have 53 minute periods). Furthermore, this extensive planning process prepares the instructor when a group discussion veers toward what the students have found relevant to their military experience—indeed, their viewpoints often surprise the civilian professor. Alternative schemas should be prepared, in order to visualize the text from several possible viewpoints, thus allowing for useful digressions. In this capacity, a civilian instructor can be useful to the institution, giving students a taste of the life that awaits them as well as proffering reflections on literary texts from a less institutional perspective. Limitations thus exist for us as instructors, yet our focus sharpens when the larger needs of the institution influence course development.

A common pedagogical danger in a service academy involves falling back on simplistic plot conversations as our students, in the main, suffer not only from a lack of time but also of sleep. One can be even more tempted to simplify course content as our courses count among the few with prominent gray areas, a tendency mirrored by our students' technological focus and military training that there is always a right and a wrong answer. Thus we in the Humanities cast doubt on mainstream thinking and encourage them to probe their own thought processes, which young people, whether in civilian or in military institutions, can be reluctant to do. Still, forcing them to react and to question aids in their overall development, as brought forth in our department's second mission goal: "To deepen and expand our students' understanding of the human condition." A conscious effort must be made to challenge their thinking, to push them to test their own ideas, and to confront their own double standards by using the course texts as a testing ground, one which will allow them to construct their own understanding of the core values of their institution more deeply and meaningfully. Only in what we deem the "fuzzy" courses is officer-development-centered discussion possible on a daily basis; the sciences, in contrast, offer opportunities to excel and to lead by doing.

An outsider may wonder, why must we teach them in the classroom what they are being taught elsewhere? This indeed was my question upon arriving at the Academy, though the answer came rapidly during our extensive new-instructor orientation. A panel of three seniors—specimens, if you will, for someone not used to seeing young people in military uniform—had been organized for a question and answer session. Initially impressed by their candor, I was soon taken aback by their cynical disparagement of the academy in front of a group of military and civilian commanders, which is what we are considered to be in the classroom. In listening to them, I realized that the classroom was one of the opportunities to chip away at their suspicion of the academy's power structure. This special type of student often embraces cynicism when watching others in the system—for example in squadron hierarchies where development, experience, and commitment are still ongoing—perform poorly or lead unconvincingly. I'll never forget the bitterness with which, three years after the fact, a student described a sophomore training officer who was making him and his fellow freshmen memorize a poem she herself did not know by heart. Since orientation, my own armor against cynicism includes living reminders of core values in everything that we do, hopefully providing students with a purpose in their readings but also with a coherent "whole" when they look back on their upper-education careers. Civilian instructors, by exhibiting behavior and setting standards congruent with cadets' military development, reinforce significantly cadets' faith in the core values of the institution; the walk led by officers thus mirrored by civilians conveys that those values extend beyond the Academy grounds and have meaning in the larger society as well.

The Academy's Mission and the Ethics of Literature

Of great importance for instructors in course preparations, therefore, is in aligning the works to be read with the guiding principles, core values, and officer training program of the institution. USAFA's core values—"Integrity First, Service before Self, and Excellence in All that We Do"—guide both cadets and faculty along their Academy journey. Supporting these, our faculty guiding principles are to "provide quality education," "promote responsibility and trust," and "prepare cadets to be career military officers." Keeping these in mind, we must also reflect upon our overall faculty mission of "educat[ing], train[ing], and inspir[ing] men and women to become officers of character, motivated to lead the United States Air Force in service to our nation" (United States Air Force 2008). Finally, our USAFA Officer Development System (ODS) sets standards for behavior year by year of a cadet's Academy career (moving by year from "Personal" to "Interpersonal," then "Teamwork" to "Organizational"). Educational outcomes are distilled from each of these larger yearly goals; all coursework planning at the department level necessarily passes through these aims. My former university, Paris XII Val de Marne, has several missions for high-ranking individuals, but not for the university itself; while there, I made my own goals for my students. Here, I must accept the larger institution's goals for my students before working toward my own aim of encouraging them to build their new identities in a constructive way.

Practically speaking, in a core English course at the Academy, incorporating foundational values may come in the form of exploring ethical and social quandaries in literary texts, for example in *Beowulf*. While the Beowulf myth presents a detailed character study in a context of strife and war, it also brings forward considerations of friendship and enmity between tribes, particularly the fragility of the former and the dire repercussions of the latter. This discussion supports several of the freshman- and sophomore-level ODS aims, such as developing polished role models, respect for human dignity, and support for ethical standards, while it evokes future ODS knowledge blocks regarding influencing and integrating the various elements with which an upper-division cadet works. For example, as Beowulf solidifies his tribe's relationship with Hrothgar's, taking men to fight another king's battle, students are led to reflect upon building a shared vision with contiguous units. After Beowulf defeats Grendel and his mother, what I call the "seam" between the larger stories pursues several independent sub-plots: a battle, a case of fratricide, and continued personal hatred between men in formerly antagonistic tribes even long after a peace has been sealed. Often dismissed in civilian schools, these episodes offer ideal reflection points regarding internecine feuds, proving that the group is not resistant to malice, just as the earlier Unferth references intimate conflict within the tribe. In Beowulf's final challenge, the disappearance of his thanes attests to real disintegration of the *comitatus*. In this case, the student is called to reflect upon the vulnerability of alliances, the need to gain a deep understanding of the enemy, and the importance of watching over the inside while dealing with the outside. What role should an intuitive officer play in each circumstance?

In my experience, from a character-study viewpoint, Shakespeare's *Julius Caesar* pairs well with USAFA's core values as well as with the ODS training. Taking Brutus as a role model, we turn the plot into an exploration of integrity, service, and excellence while also probing reason, moral courage, motivational rhetoric, and successful decision-making. The fact that he fails in most of the latter motivates our students to consider Brutus' aims and means as well as other possible courses of action, as they imagine what they would have done and determine for themselves why he did what he did. Brutus' confusion in Act II Scene 1, his composure and speech in Act III, Scene 2, and the Act IV, Scene 3 argument with Cassius all unveil the complexity behind his actions, offering students opportunities to explore the factors which figure into Brutus' decisions.

Equally instructive as a character case-study, *Hamlet*, written a year after *Julius Caesar*, allows students to pursue the revenge theme on a different level. Rather than drawing on any politico-philosophical postulates for action, as does Brutus, *Hamlet*'s eponymous hero must depend on what Leggatt has shown to be an unreliable source, the Ghost—"*What* art thou?" (Shakespeare 2006: I, i, 49-52, my emphasis)—in order to query what "to be" entails (Leggatt 2005: 55-61). Self-reflection offers Hamlet not a distorted mirror—as Cassius offers Brutus—but self-respect as well as fear of eternal damnation. As is frequently pointed out, the word "question" comes up seventeen times in the play. One *should* reflect upon murder, and Shakespeare invites his playgoers to reflect upon retributive justice carried out in the name of revenge. In discussing Hamlet's predicament with cadets, we see that the ambiguity of his situation frightens them. They construe his predicament as an impossible one, realizing that the "right" and "wrong" they are taught in other domains disappear into the fringes of a world gone wrong. What else could be closer to war? Why, we ask, couldn't Hamlet simply kill his uncle and be done with it? Why didn't Shakespeare let him? What had to happen first? Why must he feign madness in order to keep others from knowing his motives? And why is the exploration of identity so essential to understanding the play? In order to find answers, an instructor may visit "cadet land," the world of the cadet wing encased in the space of the dorms; or, if the class is academically strong enough, one may use historical or other literary examples to solicit responses. Whatever the approach, Hamlet's turmoil must become their own: why would I defend the Constitution, why would I press the trigger, and in which cases why wouldn't I? In the play, a foreign power establishes peace; what role does Hamlet's hesitation play in this reversal? Hamlet puts off deciding and allows events to resolve his quandary; clearly, an officer in today's world can do neither.

A comparison-contrast approach works well with more modern texts, such as Hemingway's *A Farewell to Arms* when Lt Frederic Henry deserts from the Italian army or the insignificance of Mike's war medals in *The Sun Also Rises*, borrowed from his tailor yet given out to girls in bars. In the first novel, tactical mistakes made by the young "Tenente" along with the growing love affair attract students' attention easily. His desertion of a cause for which he, if we can believe his words, felt anxiety—"I was always embarrassed by the words sacred, glorious, and sacrifice and the expression in vain" (1997: 169)—seems inevitable, yet what, in the changing face of the war, had led Lt Henry to this impasse? What would lead us to lay down arms and strip off signs of our officership? Why was he fighting this war in the first place? This point of course leads to that hardest of all questions, "Why are *you* here?" Or, "Why, when you saw that it wasn't what you thought, are you *still* here?" For Mike, in *The Sun Also Rises*, why are the medals worthless to him, while the tailor is so desperate to recover them? Why do they have commercial but no personal value? In fact, in that novel why does Hemingway hardly mention the war at all? Such questions must be explored in our classrooms, with ample time devoted, in Hemingway's case, to evoking what has been conspicuously omitted. As seen through these examples, institutional goals, though they make demands on our creativity, can indeed help us build stronger classes where cadets' learning from other realms of military academy life can be explored.

Though I am writing about teaching literature in military academies, one should not infer from this that only martial texts should be taught: for example, though studying the heavily martial screenplay of *Apocalypse Now* might shed light on the human condition, soldier and civilian, during the Vietnam War, reading the more abstract *M. Butterfly*, which I used for a senior seminar, may in fact supply the distance from the conflict which allows for a deeper discussion of not only East-West relations but also national culture and universal notions of human dignity. The students quickly pick up on what Hwang deems "our respective layers of cultural and sexual misperception," and understand his call "to deal with one another truthfully for our mutual good, from the common and equal ground we share as human beings" (1988: 100). For the same course, Sartre's *No Exit* brought a powerful view of human weakness into our discussions, as Estelle drowns her baby out of "cowardice" (1989: 28), while Garcin is reminded for eternity through Inez's eyes that it was his own cowardice that brought him to prey upon the frailties of his wife:

> Garcin: … how does one recognize torturers when one sees them?…
>
> Inez: They look frightened.
>
> Garcin: Frightened? But how ridiculous! Of whom should they be frightened? Of their victims?
>
> Inez: Laugh away, but I know that I'm talking about. I've often watched my face in the glass. (8-9)

What makes one courageous? What makes one a coward? How can we determine when one and not the other drives human action? Likewise, the bombs set off in *A Raisin in the Sun* offer excellent conversation opportunities regarding America's past and our operations abroad. Here I am tempted to say the cowardice of the "compact majority," to borrow Ibsen's phrase. Hansberry's play draws attention to minority empowerment as well as to economic struggle, without forgetting women's advancement into a predominantly male upper-tier labor pool. As a civilian, I can also use just such a text to broach the taboo topic of the glass ceiling in the military, perhaps in a less problematic way than a male or female officer could, due to a possible perceived bias. Our students live in a military setting, yet the works we cover needn't be limited to war literature. On the contrary, opportunities to discuss values in a civilian setting help them to rebuild their identity and explore their commitment to stay in the military.

At our institution, a text's relevance to current conflicts or world events must also be considered. Without this wider significance, a work remains limited for the students in time and scope—thus interesting to know, yet students conclude with, "well, *I* won't have to face that." For civilians, knowledge of future officers' potential challenges must be gleaned by seeking personal accounts of field experience, as relying on media reports provides only a partial picture. Understanding current difficulties in this way enables instructors to select texts appropriately. Today, the tumult during and after World War I interests the increasingly skeptical postmodern student, distrustful of institutions in general and aware of discrepancies in the rules and their application within the institution they temporarily call home. Similarly, the uncertainty and terror of Vietnam, recounted in Tobias Wolff's *In Pharaoh's Army* and Tim O'Brien's *The Things They Carried*, resonate with students' second-hand understanding of the current conflict, while the abuses as investigated in *Tiger Force* find their contemporary reflection in Abu Ghraib, albeit on a far smaller scale. Great effort is expended in discussing our own expectations of these writings as well as our shift in perception by the end of the book. Ken Bain, in *What the Best College Teachers Do*, helpfully describes the method of one successful teacher: "she carefully planned ways to challenge assumptions and put students in compelling situations in which their existing models would not work" (2004: 51). Leaving this step out would be a grave error in a military academy, as officers in the field rarely benefit from a second chance to get things right. Discussing this in an informal setting, developing careful reflexes, before sending them out to be decision-makers can, without a doubt, save lives.

Once we have set aside preconceived ideas via discussions of prior wars, we can turn to contemporary war poetry and memoirs such as Brian Turner's acclaimed collection *Here, Bullet* or John Crawford's justly angry *The Last True Story I'll Ever Tell* to illuminate the complexities of our current war on terror. In tandem with this, brief discussions of current events and news from the front lines can be woven into coursework, hopefully with the effect of getting the students to read the serious press more often and to question bias as well as recognize the often uninformed nature of blogs, a plainly unreliable contact with the civilian world. Approached in this manner, World War I, Vietnam, and the wars in Iraq and Afghanistan lend a lens through which we may analyze their fears regarding a prolonged counter-insurgency effort from a more critical vantage point. Contrastingly, and despite possibilities for exploring our "morality of altitude" in the Air Force, for our students, World War II remains something of a black-and-white documentary of what "was." Hiroshima and Nagasaki no longer seem relevant to their careers due to the type of warfare now underway. The morality of altitude often remains simply an expression we use amongst ourselves as teachers, though I have experimented with presenting the idea to students via a 1991 photograph by Ken Jarecke of a bombed vehicle, its driver's body intact though incinerated, appearing as such to be petrified in lime. This brings the reality on the ground closer to them. What works better, however, as regards developing awareness of the atrocities of war and effects on human dignity—which goes back to the pedagogical techniques I learned in France—is having them bring in their own pictures of what war means to them, pictures of what they think they will see, then have them comment on the "why" behind the picture. Their thoughtful responses never fail to impress me. In sum, we must teach them in ways that speak to them. Choosing a work simply because it is good literature may have an immediate effect, but probably not a lasting one.

As a result, some distance between the instructor and the subject matter—that is to say, not simply teaching something because one finds it to be of high value, but choosing it because of its value for the cadets, because it speaks to cadets, because it can be used to develop officers of character—can be helpful in gauging which narratives seem most pertinent. The danger of not keeping this consideration in the fore when choosing literary works is that the students will dismiss the context or refuse to entertain the content from the ethical approach desired. Thus the civilian instructor as well as the company-grade officer teaching in a service academy resembles a perpetual student due to the changing needs of the students. Not only does a seemingly protean jargon exist that one must try to master, but more importantly, knowledge of the history of warfare and artistic responses to this history must be actively pursued. My approach to bolstering my credibility with regard to the students has also included, as in France, careful listening in class, enabling me to tune into what *words* matter to cadets, what tops their individual lists of priorities as well as what elements figure into their worldviews. In tandem with this, devoting time to our Writing Center where I read student papers from across the curriculum—military strategy to philosophy, aerodynamics to Calculus III—opens the world of their studies to me. I see the questions teachers in other departments consider appropriate, the problems students are asked to solve, and in the end, I benefit from a more solid perspective on the cohesion of the Academy experience itself. Whereas in my civilian universities I could simply afford to stay up to date on my field of study and those adjacent as well as on general political and socio-economic questions in order to remain meaningfully conversant with my students, USAFA requires of an instructor an involvement with current affairs, military and non, beyond the familiar.

Coherence in curricula, multiple-level coordination with regard to aims, untiring dedication to learning: this new segment of my career has made me reflect upon my small role within the larger institution in a way I never did before. Simultaneously, I have come to view my courses as life preparation in a dramatically different way than in France. My French students, mostly first-generation college-goers, needed to learn to see themselves as others saw them, to shape that image in a positive way, and to become a curator of the storehouse of content we fed them weekly. Affording them every possible opportunity to develop themselves in a way that would enable them to get that first job was the thought in my mind at the beginning of each class. They were to go out in the world to make mistakes and grow; providing them with not only the self-assurance that they could find a good launching pad for their career but also that error was to be expected and even embraced was crucial. At the Academy, a similar vision of the students' needs would surely lead to disaster. A recent graduate described the Academy experience this way: cadets are given a feast of content that will ensure a rapid integration into the operational Air Force straight out of school, a situation not unlike my French students hoping to become an active part of the work force. From my own vantage point, however, I can say that the major difference between the two lies in the possibility to err. Our students in France needed to take on jobs where they would experience the business world, test their sense of initiative, and go forward willing to take risks, learn from mistakes, and grow. The more paternalistic style of European small business units is conducive to this type of growth. In the Air Force, however, I must remind my students that second chances come less frequently in their hierarchy, and that the higher they climb, the fewer they would find.

Hamlet: Imagination's Virtues and Limits

Now we can return to *Hamlet*. If a student who graduates has failed to meet the demands of the ODS, to develop a personal understanding of the core values, he or she compromises the larger organization. The 90/10 rule comes into play, as 90 percent of a commander's time is devoted to 10 percent of his or her subordinates (the business world rule of thumb is 80/20, which could lead one to believe there are indeed fewer problem cases in the military, though this could be a misleading comparison). We can think about Hamlet's inability to act, his fear over his own salvation, and his confusion over what we call his cause but which he himself cannot justify. Our students must, on the contrary, make the cause of defending the Constitution a personal one; to help them, we must steer their thinking from simplistic schemas toward reflexes of broader thinking. An example of a failure of this came to me, unexpectedly, in the form of a mid-semester exam in my senior theater class, wherein I asked students to describe their impressions of Edward Albee's Jerry and Peter once they had reached the end of *Zoo Story*. One answer brought me to question whether or not the student had completely missed the play's message regarding human dignity and human suffering: "Jerry—to be perfectly candid, we're not sad to see Jerry go. He was rather annoying, and was more than likely only withholding his 'zoo story' in order to maintain Peter's attention. Judging by his other stories and anecdotes, Jerry was not going to say anything particularly profound or interesting about the zoo anyway." When I read it, I felt the scathing failure of our discussions, and of our literary study in general, that had brought this student to his final semester of study without developing in him a sensitivity to literary appeals to human compassion. Had he not tried to see beyond the basic plot? Had he not seen a disenfranchised citizen in Jerry? Would he be willing to protect a Jerry at all costs? A response of this nature from a civilian student would hardly be worrisome, except that he or she would certainly fail the exam. From a senior in our major, however, the response signals a broader systemic failure.

Developing an appreciation of the human condition is of special concern in the domain of the Humanities as the development of a cadet's imagination in the context of the present war has never been as pressing. Given the United States' newly precarious economy, our degraded relations with former allies, and our struggle to export democracy to regions unfamiliar with its benefits, the capacity to conceptualize, to question, to demonstrate inventiveness, and to follow through must be promoted. As Vice Admiral James Stockdale pointed out, "Imagination is possibly the defining characteristic of the great military leader. It is the leavening from which springs what T.E. Lawrence identified on his desert battlefields as 'The unteachable, irrational, tenth'—that all-important 10 percent of combat mystery 'in which lay the test of generals'" (qtd. in Ahern and Shenk 1982: vii). To this effect, military officers are taught that their mind is their most valuable weapon. Still, we have to help students avoid falling into an absurd form of dialectical reasoning, which Bernard-Henri Lévy calls a "delirious" act wherein human suffering can in fact be rationalized through an extreme form of dialectics (2008: 67). Educating students to develop an imagination not merely for fighting evil but for developing the good with careful attention to discerning what really *is* good for humanity, constitutes one of the final stages of the ODS, and that is inspiration. Graduates must, according to the ODS, be able to influence their own and contiguous mission elements, integrate smaller units into the larger objectives, and develop others' talents while exemplifying and enforcing high standards in matters professional and moral. Added to this, when deployed they will need to seek the larger good with foreign elements as well; imagination is helpful in this, as cultural intelligence, which I taught at the graduate-studies level in France, relies not on memorizing significant visible differences—gestures, clothing, language—but on understanding others' values and in turn integrating this perception into one's decision-making processes. There is no "how-to" manual; in connection with all mission partners, determining the best means as well as the ends demands prudence, self-reflection, and an ability to contemplate the world through another's viewpoint. The students are taught to be aware of the "other," his language, his religion, his priorities, but unfortunately they often delude themselves into believing they are superior. I was recently painfully reminded of this when reading a freshman's poem, one about the "other," stating that the narrator and his other were preparing for one another, but that the narrator "would win in the end." Unlike Hamlet, this student allowed hubris to obscure his ability to imagine himself face to face with his enemy, though ostensibly this is what he set out to write. Imagination comes across in this poem, but without a clear questioning of self. By the end of their career, our students must be capable of entertaining many strategies and outcomes if they are going to become effective leaders in our twenty-first-century military. Like Hamlet, they must be reflective, realistic about their capacities, and inventive in their strategies; unlike Hamlet, they cannot postpone pressing decisions.

The selection of course texts thus constitutes one of the major considerations a Humanities instructor in a military institution will make during the year. Not only must one find ways to address questions regarding the institution's core values and officer development system, but one must also aim to create for students a simulation of their active-duty career, one that is in harmony with their own expectations yet one which at the same time pushes toward a deeper understanding of the history and current challenges of the military profession. Returning to the essence of Hamlet's query, we hope to help our students reach a level of self-reflection through which they will establish for themselves, critically and soundly, the meaning of their participation in the military. Hamlet's questioning of his convictions should become the norm, an embedded value, in this closed society of the service academy. Our first English Department mission statement articulates this aspiration in a manner that looks toward the students' effects on others: "To prepare and graduate leaders for a career in the profession of arms for the humane and just conduct of war." Literature and the faculty who teach it, civilian or military, support the ODS by creatively inspiring students to seek the good in themselves and inspire it in others. As a civilian, I have come to view this relationship as the third of Aristotle's friendship types, "friendship for the good," a virtuous relationship through which I learn, through listening and watching, how to help the students along while I offer them the opportunity to establish for themselves a new passion for being here.

Chapter 7

Teaching English at West Point: A Dialogic Narrative[1]

Karen P. Peirce and Major David C. Wood

The mission of the United States Military Academy at West Point is "To educate, train, and inspire the Corps of Cadets so that each graduate is a commissioned leader of character committed to the values of Duty, Honor, Country and prepared for a career of professional excellence and service to the Nation as an officer in the United States Army" ("West Point Mission" 2009: vi). This mission is carried out through four coexisting areas of focus, including the Academic Program, the Physical Program, the Military Program, and both formal and informal programs in moral-ethical development ("The Cadet Leader" 2007: 4). The Department of English and Philosophy at West Point falls under the Academic Program, although influences from the other program areas are felt in the classroom on a daily basis.

The key terms from this mission statement that apply most directly to the experience of teaching English at West Point are the words "educate, train, and inspire." In many ways, these concepts work well together, but at times tensions do arise between seemingly conflicting priorities. The concepts of educating and training in the military mind are inter-changeable. Army professional development schools, for example, educate leaders in the doctrine and tactics of a particular field, but in the academic sense, this is training, not educating. The lack of independent thought and the presence of an approved solution in this type of training, arguably, inhibit these schools from actually educating. The diversity of the faculty at West Point blurs this line even more. Field officers are selected to serve alongside civilian professors, and what these field officers consider educating differs, sometimes greatly, from the beliefs of professional educators. For example, West Point's Department of English and Philosophy advocates mandatory student-teacher conferences occasionally each term. Classes are canceled for several days so that cadets can meet with faculty to discuss an assignment in progress. Students are expected to report to the conferences in a timely manner, come prepared with the required materials, and stay for the allotted time. Such an activity certainly trains cadets to follow directions. However, to educate cadets to take responsibility for their own learning, some faculty would prefer to make these conferences optional.

1 The views expressed herein are those of the authors and do not reflect the position of the United States Military Academy, the Department of the Army, or the Department of Defense.

These faculty members, civilian and military alike, feel that students need to learn that consequences exist for their choices, and they feel students should learn from these consequences as they advance through their education. This situation is just one example of some of the tensions that can arise between the key terms embedded in West Point's mission statement.

We invite you to join us as we explore tensions such as these in more detail. Written jointly by one civilian and one military member of the Department of English and Philosophy faculty, this chapter represents the viewpoints of these two authors through a dialogic narrative, thus giving equal weight to both voices as we explore our personal opinions on these contentious issues.

Karen: First, I'd like to take a moment to introduce myself. For three and a half years, from 2006-2009, I was a member of the civilian faculty appointed as Assistant Professor in the Department of English and Philosophy at West Point. I have a PhD in Rhetoric, Composition, and the Teaching of English from the University of Arizona, and during my years at West Point I was at times one of three and at times one of two PhD faculty there with a degree in rhetoric and composition. My dissertation focused on intercultural rhetoric, aiming to find ways for people of diverse cultural backgrounds to communicate so as to reach mutual understanding. I suppose this research interest is what drew me to West Point, for it was an opportunity for me to experience a slice of military culture. I will confess to experiencing not a small amount of culture shock in the three years that I worked there, and this chapter will, among other things, shed some light on the differences between teaching at a military institution as compared to teaching at a civilian college or university. Before we delve into those issues, though, you should meet the other author of this chapter.

Dave: "GROUP, ATTENTION!" Perhaps that fits the stereotype too strongly, so maybe I should begin by introducing myself. I am a member of the rotating military faculty in the Department of English and Philosophy at West Point, and I have served as an instructor for two years and an assistant professor in my last year. My service to the nation spans twelve years and one combat deployment to Iraq. The Army is my life and always has been as I grew up as a third generation career soldier. After much trepidation concerning expected career paths, I pursued the opportunity to teach at West Point. My selection by the Department Head allowed me to pursue a Master's of English (Specialization in Nineteenth-Century Studies) from the University of Nebraska-Lincoln (UNL). As with many of my military colleagues at West Point, my studies focused primarily on areas related to what I thought I would teach during my tour of duty at West Point. UNL offered me opportunities to study literature and, most importantly, an opportunity to teach and study composition.

As Karen stated, I too experienced my own forms of culture shock both while attending UNL and at West Point. My transition to UNL was one year removed from my deployment to Iraq. I found myself fully immersed into the civilian world because of the lack of military presence in Lincoln, Nebraska. While looking at my graduate student peers and my students with a new set of eyes, I found that the wars in Iraq and Afghanistan were not a part of their daily lives. They found me as much an oddity as I found them, but much to all of our surprise, we found each other to be "normal."

It was interesting to discover that our mutual stereotypes of each other were inaccurate. As an example, one day in class, my students and I discussed the recent fascination with naming generations. Many of my students responded negatively to the current naming conventions for their generation. They did not feel a connection to Jean Twenge's "Generation Me" label or the frequently used "iGeneration." The students responded to the idea of naming their generation by stating that their generation has not done anything yet. I found myself explaining to these students that their generation is accomplishing great work already. They were surprised to see some parallels to the Greatest Generation. I used vignettes from my combat experiences and worked to remove the perceived wall of differences between those 18-20 year olds that go directly to college and those that serve the nation. I think they were willing to listen to me because of my presence in the classroom and how I did not hide the fact that I was a career military officer. Surprisingly enough, my interactions with my peers and department faculty members also worked to disrupt stereotypes.

In many ways, I think my peers found my presence to be something more than a novel experience. They found my self-reflective considerations of the identity juggling act that I was performing to be intriguing. For the most part, my interactions with faculty members were cordial and respectful. Only once did I receive a recommendation to steer clear of a particular instructor in my studies. These interactions with peers and faculty members resulted in a back-and-forth discourse about identity. I found this search for a complete understanding of my changing identity to be a time-consuming but necessary part of this opportunity to be a student, teacher, and career military officer.

My peers and I would frequently discuss our two differing worlds. We found similarities and differences. We found differing perspectives and began to see the other side's reality. Frequently, we also would discuss a teacher's presence in the classroom and the problems with authority in the classroom. I frequently found these conversations to be useful in how I was crafting my teaching persona. My peers would be amazed that the students did not hesitate to use my military title. In fact, for many of my peers and my students, it was their first contact with someone who has earned a title based on their work outside of an educational setting. It was also for many the first time that they had interacted with someone who has served in the Army as long as I have. I found these interactions to be also helpful in reflecting on my days as a company commander where I acted as the primary teacher and trainer for my platoon leaders who were just a few years older than my students.

As I transitioned to West Point, I found that the students here were all too willing to celebrate the great opportunities for their generation to shape the nation and the world. I found the students to have a balanced sense of optimism and realism about the world. I found their curiosity about my career to be insatiable. I also found their sense of familiarity with me to be somewhat shocking. In the Army, majors are feared, and with good reason too, considering my past experiences, but the cadets interact with captains and majors frequently here. Comically, I find it interesting when I remind my cadets that I was just a student a few years ago. This reminder jolts the cadets as a lesson that their education does not stop here; it is a reminder that the Army encourages the educational growth of its officers. As I continued to adjust to life at West Point, my peers still had many of the concerns about authority and presence in the classroom. In this case, it was more a concern based on the instructors' backgrounds. My civilian colleagues worried about having enough authority. My military colleagues worried about establishing too much authority and thus leading the students to believe that "an approved solution" exists in the worlds of composition and literature. I found myself continuing my juggling act of all the components of my identity. Karen, I'd like to hear more about the culture shock you experienced. Did you feel some trepidation about coming to teach at West Point?

Karen: Honestly, no. Instead I felt excitement at immersing myself in a new cultural experience. During the job interview process, everyone I dealt with was polite, professional, and welcoming, but there were signs that I was observing a different culture at work. For example, during my campus visit, rather than taking an elevator to get to a sixth floor office, the officer escorting me and I walked up the six flights of stairs. Most people in civilian society would generally take the elevator rather than climb more than one or two flights, but the officer did not hesitate to start up the stairs instead. This, I surmised, was due to the cultural norm in the Army of showing physical fitness, and it was an experience that hinted at the differences that exist between military and civilian cultures. I realized I had a lot of learning to do about this military culture, and I was excited about that prospect. Before I worked at West Point, I had no previous experience with the military beyond what I had seen in the media. My only images of soldiers were from war movies where there was a lot of blood and gore, or news reports about soldiers conducting combat missions and raids on houses in Iraq. I also had seen soldiers who toted machine guns patrolling airports in the months following 9/11. All of these images are of the intimidating sort, but they clashed with my interview experience at West Point, so I was curious to learn more about this apparently complex culture.

During the interview process, the military faculty I interacted with wore their Class As or Class Bs, the sorts of uniforms we might see high ranking officers wear when giving testimony before Congress, and they were composed and polite, not at all fitting the image of violence I associated with images of soldiers wearing their combat uniforms. I did find myself feeling nervous when I reported to the local Air Force base in Tucson to arrange my government-funded move to West Point, as the gate guards and all the military personnel I saw there were wearing combat uniforms. In fact, it took me several months once I arrived at West Point to get used to seeing my colleagues and students sometimes wearing combat uniforms to class. At first I felt quite uncomfortable about it, because the negative connotations that particular uniform conjured up for me contradicted my experiences with the faculty and students wearing it. But over time, I came to realize that to the people wearing the uniform, it was just another set of clothes to put on as opposed to standing for violence and destruction. I realized that the officers and cadets behind those clothes were still the same people on the days they wore the combat uniforms as they were when they wore their more suit-like Class B or A uniforms. So that intimidation factor associated with the combat uniform disappeared over time and the excitement I felt about this new cultural exploration prevailed.

I did notice over time, though, that some of the authority issues you mention did arise for me, perhaps more than they had in my previous teaching experiences. In my civilian teaching experiences, in general I felt that students and fellow faculty treated me with respect, and I was afforded plenty of opportunity to apply my specialized knowledge as a leader for both students and colleagues. At West Point, however, I sometimes wondered if my civilian status acted as a hindrance to me. In the realm of teaching, for example, often the faculty discussed our classroom experiences, and several of the military officers were surprised to hear me tell of disciplinary issues that sometimes came up in my classroom because they were not experiencing the same in theirs. I told them about excuses for late papers, back talk in response to directions, grumbling about grades, all things I had experienced in other educational settings, but which my military colleagues reported they were not experiencing. I found myself wondering whether I wanted my students to be more polite and less forthcoming about their feelings as they were for the military faculty, or if I liked the idea that my classroom provided my students with an outlet they might not otherwise have. I also found myself wondering if any of these issues related to me being a woman, a minority population at West Point making up only about 15 percent of the students and faculty. While gender dynamics are always at work in any classroom, certainly they seemed more pronounced at West Point. Was I providing an invaluable lesson to my students in the ability to work for a woman boss/leader? Or, particularly with the more advanced students I taught, was their education suffering because they had been conditioned for so long to respond to male leadership that they had become somewhat unable to learn from a woman?

One of the first manifestations of these questions that I noticed when I began teaching at West Point was classroom performance, especially in comparison to my previous teaching experiences. One of the courses I taught most at the University of Arizona was Advanced First-Year Composition, also known as Honors Composition. The students I taught in that class were high-achievers in high school English, earning a 4 or 5 on the AP English exam. They were motivated learners, almost always completing all the assigned homework, looking for challenging ways to interpret the requirements of the assigned papers, and participating freely and with deep insight in class conversations. When I accepted the position at West

Point, knowing that West Point has a reputation as one of the country's premiere educational institutions, I expected all of my classes would be filled with students like the Honors Composition students I taught in Arizona. What I quickly realized is that teaching students in an honors English track is not the same as teaching students who are high performers overall, but not necessarily tracked for English. While my students at West Point may have had higher composite SAT scores, may have been more likely to be student class president while in high school, may have been more likely to letter in multiple sports than my honors composition students in Arizona, English may not have been the favored subject for the cadets in my mixed-ability English classes at West Point. This affects motivation for completing homework, motivation for taking essay assignments beyond the bare minimum requirements, and motivation for participating in class discussions.

Further complicating the situation of classroom behavior is the extent to which my students at West Point knew each other outside of the classroom. At the University of Arizona, where incoming classes of first-year students regularly reached sizes of 6,000 students or more, students enjoyed a certain amount of anonymity in the classroom. They could be the geeky English type within the class hour, and then leave the classroom, never to fear that being eager to participate in discussion or being sure to have completed all the homework would become known by the larger student body. At West Point, however, where each incoming plebe class is more around the number of 1,300 students, and where all students live in the barracks, it can be harder to hide one's English geekiness outside of the classroom. Therefore, I witnessed tickling, note-passing, and other disruptive behavior in my West Point English classes to an extent that I hadn't seen since I taught 9th grade English many years ago. Admittedly, this came as a great surprise to me when I first entered the classroom. Have you witnessed anything like this?

Dave: I have not had any more annoyances than what I experienced teaching at the University of Nebraska, but it does concern me even more when I do because of the ever-present environment of military decorum and regulation. I find it difficult at times to balance creating a classroom that celebrates and encourages individual thought with a classroom that requires some modicum of military professionalism. This lack of professionalism is possibly not all that different from the civilian university. It is best seen in the student who speaks without thinking or the cadets who talk while I teach. These moments, though, are anomalies that chafe against the stereotypical quiet, respectful, hand-raising atmosphere of the military's training schools.

I find my classroom to be a fine balance of attempting not to stifle individual thought and attempting to teach students to express their individual thoughts in a polite manner. I enjoy the occasional lapses into playful banter, possibly because it resembles so much of our Army lives, but I also do not tolerate playful banter that becomes unfocused or disrespectful. This balance proves even more difficult when students (both first and third year) care only if the subject material is what they consider to be worthy of placing in their "rucksack" for military life. I can sympathize with these students on some texts or ideas because of the seemingly indirect connection to their future military careers. Yes, it does come as a surprise that our cadets come not only with physical differences, but a difference in how they approach the classroom. Some are tomorrow's future generals, while others will become leading researchers in their fields, and finally, some will become leaders of industry; I can see the need to classify their teachings into how they relate to their perceptions of what their life may need in two to four years. Overall, though, I have the benefit of hindsight to see how what I am teaching them today will better prepare them for not only the leadership challenges of tomorrow, but also to place them in an advantageous position for continued success in the classroom.

My only other surprising gripe is the reliance on "proven" essay writing techniques. Surprising for their academic pedigree, it is rare to find a student who is willing to re-see their writing process or engage their perceptions beyond the so-called "approved solution." This lack of engagement is perhaps manageable in an atmosphere that emphasizes calculated risk-taking, but many of these students, especially by the third year, have become somewhat risk-adverse in classes that do not seem to relate to their major. At times and possibly an effect of the "approved solution" seeking, these students will become more resistant to a logical connection they did not see or to a writing process that while unfamiliar seems to hold greater potential for success. I found it surprising at West Point because I did not see it in my civilian students. It is also surprising in light of how we emphasize risk while teaching military tactics or while encouraging a competitive culture in both academics and athletics where risk, be it physical, mental, or emotional, is inherent because those environments place value on calculated risk and being flexible to changing conditions.

Despite those concerns, I do see more students here embrace not only the idea of joining the military but the idea of joining the academic community. I frequently see students who are intrigued with the possibilities of research or even just exploring their personal connection to a subject. When they find that "desire" for a subject, I see their rapt desire to learn more, and it impresses on me how they might perform as military leaders. These students are not the norm, however, and suffice it to say, with their schedules, it is even more impressive when the "average" does show that higher level of interest. This higher level of interest, though, does not always translate into success.

This contradiction of higher interest and lack of success can be seen in both the plebe (freshmen) and cow (junior) composition classrooms. In EN302 specifically, our Advanced Composition course for juniors, we have a greater opportunity to draw upon the wide array of knowledge and interests of our students based on their previous two years of learning, but it is at this point where many of the cadets have begun their coursework for their majors. Most students, based on the successes and failures of previous semesters, have decided whether or not they can write well, and that self-assessment leads to mediocre performance and, even more worrisome, a troublesome lack of effort. Additionally, the course also requires a cultural studies component or what we like to call a "cultural destination." I have found this two-headed complication to provide moments of great intellectual inspiration for many of my students as they not only explore their writing but a different culture, but it is also fraught with the negative response of why study a place such as Japan in the midst of a writing course? For me as for many of our peers, EN302 proves to be the most challenging course to teach because of this juxtaposition of student reactions and ambitious course objectives.

As their civilian peers do, many of the cadets make conscious decisions as to what coursework will take priority. In EN302, those who are excited by the opportunities involved in such a complex course put it at the top of their priority list. On the other hand, those who have already decided they are not writers and are further overwhelmed by the cultural destination aspect of the course put the class at the bottom of their priority list. Furthermore, many of their decisions are classic examples of Pavlov's theory of conditioning. In some classes, students are conditioned to respond to points-based assessments, so these students will study hard the night before for a 50 point "exam" in a 2,500 point course rather than working on a 150 point essay in a 1,200 point course. At times, I can see the merits in this Pavlovian approach. It is actually a very military approach to teaching, or should I say, training. It teaches, it conditions, it asks for immediate feedback on the course material, and assesses whether or not the students grasp the material. The student learns through the assessment if they have had enough recall to move on to the next topic or if they need to conduct Additional Instruction (AI), what other campuses might call extra help, to reach some level of "mastery." This approach also subtly relies on the students' need for immediate affirmation of their efforts in the constant swells of the multifaceted academic, military, and physical life of a cadet.

Karen: I think I understand what you mean when you comment on the cadet need for positive affirmation. Perhaps in a place that does put so much stress on perfect parade uniforms, and being absolutely punctual, and early morning inspections of rooms, and where punishments are meted out for shortcomings, it is understandable that cadets would want to play it safe, so to speak, in all areas of their lives, rather than taking risks that could go wrong. In other words, in a place so focused on military training to get things right, cadets may have a hard time making the transition between this sort of training into an academic mindset where risk-taking is rewarded and attempts to be clever or original are lauded, even if the results are not perfect.

And your comments about cadets prioritizing the demands on their time resonate with my experiences with the cadets as well. I understand the military exigency to condition people to operate under high levels of stress, little sleep, and less than ideal conditions. But I often found myself wondering if this conditioning is meant to result in the best performance, or just passable performance. It seems to me that perhaps the aim might be to achieve the best possible performance under less than ideal circumstances, and this conclusion made me sorrowful at times over the possibly lost academic potential in some of the brightest cadets I encountered. I wondered to myself at times what sort of academic work the brightest cadets in my classes might have been able to produce if they were in different circumstances, if they weren't spending their weekends toting guns around training fields, or attending mandatory sporting events, or reading maps to find their way in the woods. And when I mourned over the possible reduction in depth of thought I saw in the top performers' essays, I wondered how these stresses affected the work of my average or below average students, and I started to wonder if I should be grading on a different scale to give these students' GPA a break.

Having said all that, however, I also occasionally caught myself. Was I over-romanticizing the non-military academy student's life? What about the civilian students who work full time, are single parents, take care of their own aging parents, live in their cars, or face any number of other obstacles keeping them from performing under ideal conditions? When I recognized that time management and juggling competing demands are life skills gained in the midst of any education, not just one at West Point, I became less likely to feel guilty for awarding the grade I thought any given cadet's work deserved.

Dave: Karen, this line of discussion brings us to an important and possibly surprising aspect of West Point: how West Point mitigates coursework failure with its mission to produce leaders of character. I am always struck by the seemingly complex scheduling that the cadets attempt to balance in order to succeed in their courses and how they portray that busy-ness. As a former company commander and leader in combat, I can see the merits in a system that taxes even the best student's time management skills, but I also find that this all-consuming 47 month experience leaves little time for daily critical introspection and reflection, which is crucial to developing a self-evolving leader.

I had a chuckle with your comment about the "gun-toting weekends." Many of the cadets feel that their military training falls short of what they perceive as needed. In fact, many of the cadets receive only one weekend training event a semester as a supplement to their 55 minute military science and leadership courses which occur as normally scheduled classes alongside their other coursework. Due to these training constraints, many of the cadets do not conduct any serious training until the summer. The cadets do like to make mention of how their "military" activities such as formations, inspections, sports, and a variety of other events add additional burdens to their ability to complete their homework, though. Although dissimilar from the variety of "military" activities they will face in their units or in combat, these unique stressors act to simulate the real-world minutiae of stressors such as patrols, training plans, maintenance problems, and leadership challenges they will have to manage in their day-to-day activities.

In order to help the cadets manage their time at West Point and thus, reduce the specter of failure, the instructional departments adhere to an institutional rule of two hours of homework for every one hour in class. The student schedule from 0600-1930 (7:30PM) is dictated, and the cadets do have a required Evening Study Period (ESP) from 1930-2330 (7:30PM-11:30PM), but the cadets can make their decisions as to what subjects they will study during this evening time. ESP provides a means to balance their coursework and possibly as a saving grace from failing a particular course. But does ESP actually allow students to succeed, or is it just another means of scheduling the cadet's time? Does it take the responsibility away from the students to manage their time? ESP is all of these aspects. I find it an interesting way to mitigate failure in the cadets' coursework. What other school requires a specified student study period?

Another means that West Point uses to encourage responsibility that seems unique in comparison to my civilian teaching experience is that, at least in the Department of English and Philosophy, the cadets are required to submit an assignment no matter how late and even if the assignment is so late that the instructor has no choice but to receive Course Director permission for a zero percent "F" on the assignment. Additionally, instructors are required to submit a Cadet Observation Report (COR) for any and all late assignments and any assignment receiving a failing grade. The COR is an electronic message sent to the cadet's chain of command that addresses non-normative behavior such as outstanding positive actions or exceptionally negative performance. These measures are seemingly contradictory in terms of personal accountability and consequence.

From a military perspective, it may give these future officers a worrisome impression of a permanent safety net instead of placing them in the role of being fully responsible for their actions and experienced in failure. Despite this concern, the COR acts as a corrective used by the cadet's chain of command to ensure the cadet's responsibility, a measure much more hands-on than what their civilian peers may face. For in the military world, a failure to complete an assignment might result in a counseling session and finally a completion of the assignment. It may result in a verbal reprimand. It may also result in a relief of duties which will result in a negative evaluation. The corrective measure is in theory supposed to match the level of importance of that assignment. The COR achieves this corrective measure, but since the correction takes place in the cadet's company outside of the instructor's purview, the instructor-student relationship enters a complex state of responsibility and culpability. Karen, what do you see as some of the measures in the classroom that West Point uses to mitigate or to inadvertently create failure?

Karen: I suppose another aspect of life at West Point that might be related to this topic of academic success and failure is Additional Instruction (AI), which you mentioned earlier. I find it interesting that at West Point faculty members are not allowed to set office hours as they do at other institutions. Instead, faculty members are expected to be "on call," available to meet with students at any time during the academic day, which runs from 7:30 in the morning through 4:00 in the afternoon. It is not an unusual circumstance to find a faculty member who does not teach until 2:00 in the afternoon, therefore, in the office at 7:30 in the morning because that is the only time the cadet has a free hour to meet for extra help. Furthermore, while it may be tempting to at times turn down such requests in order to gain a few hours of precious sleep, I, personally, never felt free to do so for fear of being considered delinquent in my efforts to keep a cadet from possibly experiencing academic difficulty. This is because at the end of each term, any student who fails a class has his or her record reviewed first by the Course Director, then the Department Head, then the Class Committee (a committee consisting of all the department heads and the Dean) that reviews each cadet for retention at West Point, and finally the Academic Board (a committee of selected department heads, the Dean, and the Superintendent) that reviews each cadet recommended by the Class Committee for separation from West Point. One of the elements required for these reviews is a record of AI received by that cadet in each course. The implication as I understood it is that if I did not comply with requests for AI and that cadet should eventually fail the course, my having provided AI or not having provided AI would become an item of discussion. To me, such scrutiny of the faculty member's extra efforts outside of class to tutor individual cadets as needed seems to put undue pressure on the faculty member to ensure the student does not fail.

The faculty is told that West Point is one of the finest, if not the finest, academic institution in the country, as evidenced by rankings in *Forbes Magazine* and *US News and World Report*, because of the dedicated faculty and their effort to ensure that cadets make it through the 47 month experience. However, I was often left wondering what happened to the student's individual responsibility for success. Perhaps if I were to attend the briefing by the Dean to the cadets, instead of attending his brief to the faculty, I would hear him exhorting the students to do what it takes to get themselves through their courses. Instead, what I heard was faculty members being expected to work harder and longer to help any student who needed it at any time.

I suppose another related element to this line of discussion is the remedial approach the composition courses at West Point take with students who are basic writers. In writing programs around the country, students are regularly placed into some sort of basic writing course if they are not deemed capable of the standard writing track. In fact, more and more often these placements are determined through a system of directed self-placement (DSP). In these systems, the institution makes recommendations to students regarding their placement in writing courses, often based on high school GPA, SAT or ACT scores, number of honors courses taken, AP scores, and so on, and the students either accept the recommendation or make a case for changing the placements, often by submitting a portfolio of writing as evidence for why the placement should be changed. I worked in the University of Arizona Writing Program during its transition to a DSP system and found it heartening to see students take an active role in determining the path of their academic careers. We were also developing a new basic writing course at the time, a course that mirrored the standard course curriculum but provided extra support services through the writing center, and it was interesting to see the number of students who requested a change to their placement out of the standard course and into this basic writing course.

However, no such system for placement into writing courses exists at West Point. In fact, only one writing course is offered for all first year cadets, EN101.

Some students, approximately 90 out of the greater than 1,000 incoming first-year students, place out of this course and into the first-year literature course instead. Placement is determined via a timed essay test taken in the midst of summer military training that is read and evaluated by literature faculty. Other students place out of the first year writing course and into an ESL program offered by the Department of Foreign Languages, but the only students allowed to take part in the verbal and written evaluation process for this program are international cadets. All cadets other than these few take the only writing course offered to first-year students, EN101.

Because no placement measure is used to determine at-risk students, inevitably some students are unable to perform to the expected standards in EN101 and fail the course. When this happens, they are enrolled in EN101 for a second time. This course is commonly called EN101R, the R standing for "remedial," although no such designation appears on the student's transcript, and the grade received in the course the second time replaces the grade earned the first time. I taught this course once, and in my planning I was curious what would happen if a cadet should fail the course a second time. I was told "no cadets ever fail EN101R," and I was never sure whether this was meant to be a joke or not. In any case, one student did fail the course, despite my multiple offers for AI appointments, mainly due to late penalties on his work. All it takes in the Department of English and Philosophy is 24 hours of tardiness on a major assignment for it to receive an F, after all, and with multiple major assignments late more than 24 hours, this poor student had no chance, really. It is West Point policy that a student who fails the same academic course twice in one academic year is separated, which is known elsewhere as being expelled, and ultimately this was the student's fate, despite multiple emails and phone calls from his mother to the Academic Board attempting to place the blame on me and reverse the decision.

What strikes me is the sort of double messages for students that I read in these tales. On the one hand, they are led to expect that faculty are at their bidding to help them through the coursework through AI on demand, a message of generosity at best and perhaps at worst a message that effort alone will suffice to earn a passing grade. On the other hand, they are told there are no exceptions, no tracks for lower performers (at least in the Department of English and Philosophy), so they had better be able to make it at the average standard, or they are not good enough, a policy that has a harsher tone than the on-demand AI message. Next, though, they are given a second chance in a remedial course, a seemingly generous offer, but if they don't make it through, the standard operating procedure is dismissal from USMA, a rather cold shoulder. Out of all this, I find it hard to read what the ultimate message is about failure at West Point. Perhaps faculty and students are expected to do whatever it takes to avoid it, putting the burden on the individuals involved, even though the systems set up, such as the courses offered and penalties for failure, seem to offer little admission that individual learning styles differ and that multiple attempts may need to be made to reach success. What do you think about these issues, Dave?

Dave: I have found failure to be much more of a concern here at West Point. In some ways, it makes more sense because of the mission of West Point. We are here to educate and train future leaders of character. We have a responsibility to the American people to provide the best education we can; we have a responsibility to the mothers and fathers whose children serve as enlisted soldiers in uniform to provide them with the best leaders we can. Many of our practices to mitigate failure possibly arise because of these responsibilities and the intersections between tradition, the blending of education and training, and the needs of the Army.

In many ways, today's realities of war place this responsibility on the instructor and on the means of evaluating a student's intellectual capabilities and demonstrated leadership potential. When those days occur that West Point buries one of its own in its historic graveyard, the reality of the world into which we send our students, our cadets, resonates with each of us. No longer is the question hypothetical as to how this young leader might perform under combat. When asked by the cadets as to my expectations of them as students or my expectations of subordinates, I know that these individuals are cultivating a sense of who they are and who they will be. We judge and evaluate with a more critical eye as to their potential for further responsibility; we judge and evaluate the best we can based on our combined experiences of teaching, training, and leading. This additional responsibility of all the many types of instructors, be they civilian or military, at West Point leads to pedagogical tensions over who should gain admission, who should continue on when they have failed, and finally who should be commissioned. Truly, a one of the kind experience that I never imagined while in my classroom at UNL.

Another aspect of our conversation about our techniques not only to mitigate failure but also to breed success in this demanding system is the role of West Point's primary pedagogical methodology: the Thayer Method. Named for West Point's most beloved superintendent of yore, Sylvanus Thayer, the method comprises multiple aspects, the first of which is smaller class sizes. For example, class sizes are capped at 18 in the First-year and Advanced Composition courses. Second, cadets are expected to come to class each day prepared either to recite detailed and accurate explanations or to enact the homework material. In all courses at West Point, teaching is based on discussions, board work, or practical exercises, thus limiting the anonymity of lecture-based courses found in other institutions. Since 1817, the Thayer Method has evolved from its initial focus on math, engineering, and French to support a variety of courses.

Vestiges of the Thayer Method still exist in our core composition courses, but generally not as a sole means of teaching. One factor that continues this methodology is the shorter class periods of 55 minutes two to three times a week as opposed to most civilian institutions that offer 80-90 minute sessions twice a week. Compounding this shorter class time is that students generally have four to six classes every day. It is no wonder that West Point mandated an evening study period, for if you have two hours of homework for each class session the following day, it is frequently impossible to complete all of the required work without effective time management skills.

Needless to say, the cadets both love and hate the Thayer Method. Many enjoy the sense of responsibility that the pedagogy espouses, while others find that even with the best of efforts, their results achieve only a satisfactory grade rather than the lofty grades that I have seen and given in my teaching in a civilian university. Yes, some cadets feel this system places them at a disadvantage, especially if they have never encountered a pedagogy such as this one. And yet, many cadets do have success with this system; some cadets even manage their time well enough to excel spectacularly, as evidenced by the high number of Fulbright, Rhodes, Truman, and other scholarship winners West Point has had over the last twenty years.

Karen: You're right, Dave. Your discussion of the Thayer Method serves as a reminder of the unique mission of the military academies in our country. To go back to our introduction, the primary goal of West Point is to commission officers into the United States Army. A well-rounded college education can certainly be a part of that goal, but it is not the entire goal. At West Point, a pronounced emphasis exists on the need for cadets to perform under stress, and although this might have costs in the classroom, it is most likely absolutely necessary to prepare these students for their future roles as officers, especially in a time of combat. As with any other institution, the standard operating procedure at West Point has benefits and costs. The students who have chosen to follow this path have ample time to change direction if they are dissatisfied with their experience, as they are not bound to provide service to the country in the form of active duty until the first day of their third year at West Point. Should they decide before then that the costs outweigh the benefits, they can transfer out of West Point without owing any military service to the nation. However, the vast majority of cadets choose to stay for the full 47 month experience that comprises a West Point diploma and commission, and if the benefits outweigh the costs for them, ultimately perhaps that is all that matters.

Dave: Despite the challenges of attending West Point, I am glad so many cadets do complete their 47 month experience at West Point and go on to become officers in the Army. Even for myself, I consider my time at West Point as having been uniquely rewarding. The opportunity to teach, mentor, and train this select group of individuals has been extraordinary. I would never have imagined the seemingly conflicting natures of this institution, but I have confidence in our unique teaching and training environment. West Point not only gives these students a chance to become leaders of character, it also has given me a chance to become a better teacher and leader myself. I know my experiences here will continue to aid me as I transition back to the "muddy boots" world of deployments and combat operations.

Chapter 8
Cocked and Ready: The Humanities and Homosociality at the Royal Military College of Canada

Huw Osborne

Introduction

In June 2009, the US Supreme Court refused to hear Capt. James Pietrangelo's appeal that argued that the US military's 1993 "Don't Ask, Don't Tell" policy was unconstitutional. As I write, about one month after this event, neither the Supreme Court nor the Obama administration has commented on the decision. Recently, Senator Kirsten Gillibrand succeeded in getting the Senate Armed Services Committee to hold hearings on "Don't Ask, Don't Tell," but this triumph must be qualified by the fact that it is a back-up plan to the failure of Gillibrand's proposed amendment to the Military Reauthorization Act that would have stopped all formal investigations into service men and women's sexualities. At the moment, the culture of silence on homosexuality in the US military is essentially intact.[1]

According to Judith Butler, this silence speaks volumes. Butler argues that the US military's prohibition of the utterance "I am a homosexual" places the discursive power of homosexuality at the center of military self-definition (Butler 2005). The US military defines the speaking of one's homosexuality as a homosexual act, so that speaking and doing (and speaking to and doing to) become one and the same, thereby transmitting sexuality through speech. However, by acknowledging the power of such "performative speech,"[2] the military is caught in its own dependence upon the power of the perversion in establishing and preserving the sanctity of

1 In early 2010, at the behest of the Obama administration, the US military has begun the process of dismantling "Don't Ask, Don't Tell." It is not yet clear what will replace this policy, but, as the Canadian example discussed below illustrates, policy change is only the first step in addressing the entrenched cultural resistance to homosexuality in the military.

2 Judith Butler defines performative speech in the following terms: "Performative acts are forms of authoritative speech: most performatives, for instance, are statements that, in the uttering, also perform a certain action and exercise a binding power. Implicated in a network of authorization and punishment, performatives tend to include legal sentences, baptisms, inaugurations, declarations of ownership, statements which not only perform an action, but confer a binding power on the action performed" (Butler 1993: 225).

male homosociality. That is to say, "social cohesion in the military requires the prohibition on homosexuality, where that cohesion was then described as a magical *je ne sais quoi* that kept military men glued together. The formulation might read: *We must not have our homosexuality in order to have our homosexuality: please take it/don't take it away from us*" (Butler 2005: 146). Essentially, the military needs to deploy the threat of homosexuality in order to perform its banishment, which maintains the speech-performance-act of homosexuality at the heart of masculine military culture.

This chapter is about my students' reactions to homophobic and homoerotic homosociality at the Royal Military College of Canada (RMC) as it manifested itself in the literature classroom. Of course, Canada has no such "Don't Ask, Don't Tell" policy. The Canadian military granted equal rights to gay, lesbian and bisexual servicemen and women in 1992 in complete compliance with The Charter of Rights and Freedoms and the Canadian Human Rights Act. Further, in 2004, just after my first year of teaching at the Royal Military College of Canada, one of my students married his partner, becoming the first man in the Canadian military to marry another man.

Despite this difference in the official policies of the Canadian and US militaries, it may be argued that the paranoia over speaking homosexuality is very much the same across American, Canadian, and Australian military academies. In terms similar to Butler's, Carol Burke convincingly demonstrates that the recognition of, and control over, the performative power of speech is especially significant in the intensely ritualized environment of military academies, where performing and becoming are so closely tied together, and where ceremonial pronouncements, spoken orders, and ritualized speech-acts so often define one's presence within the masculine community. Whereas Butler exposes the centrality of homosexuality to the legal preservation of heterosexual male-bonding, Burke demonstrates why that heterosexual male-bonding is so haunted by the threat of homosexuality. Indeed, as Burke explains, in a world where actual sex is prohibited, speech becomes the primary means of sexual intercourse between men:

> Talking about sex is more important than performing the act, because sexual banter welcomes one's mates as co-participants, cementing fraternal bonds with shared homoerotic fervor—homoerotic not because they talk only about homosexual sex, which in fact they do a good deal, but because even their talk of heterosexual encounters becomes in the retelling a shared erotic experience in the all-male group. The cadet rarely recounts sexual exploits with someone he truly cares about; rather, he shares with the group intimate details (both real and imagined) of encounters with women he has no attraction to, women he speaks of as if they were disposable. In this way, heterosexual sex is simply the pretext for fraternal bonding, for the homosocial. (Burke 2004: 117)[3]

3 An RMC parallel to Burke's analysis of homosocial sexual bonding is found in the Eighth Squadron "Chicken and Porn" party, in which male cadets gather in one or two

According to Burke, as a form of male-bonding, sexual performance in speech takes primacy over the act of sex. Like Butler, Burke argues that the line between speaking and being is very fine indeed. One can very easily imagine the paranoid need to maintain a very fierce control over the use of the term "homosexual," so that the already transgressed lines of, on the one hand, speaking and being, and, on the other hand, homosociality, homoeroticism, and homosexuality do not appear quite as irrevocably entwined as they clearly are.[4]

I begin with this discussion of "speaking sexuality" in the military and military academic context because of its relation to the literary studies classroom, where words and modes of expression are our material and where discussions through the spoken word are our tools. In such a space, the potential for transgression is severe. This may be one reason for a common resistance to literature among many cadets. In any given year, the English program at RMC has an average of 20 majors, which is 1.6 percent of the on-campus student population, and it generally graduates about four honors students each year, which is 0.3 percent of the on-campus student population. These numbers, however, do not tell us anything about the cultural and ideological resistance to English literature in military academy culture. Some of this resistance may derive from the fact that literature rarely provides easy answers. It tends to open up areas of possibility rather than provide reassuring values to a community built upon shared sense of belonging. Both Elizabeth D. Samet at the US Military Academy and Bruce Fleming at the US Naval Academy have encountered this expectation that literature for future officers be motivational (Samet 2007: 38-39, Fleming 2005: 34-38). Literature must help men serve in combat through celebrating such values as heroism, loyalty, honor, and obedience. Generally speaking, my students seem to approach literature, especially poetry, as irrelevant to the practical activities of making war; it is something to be tolerated and humored rather than worked at with constructive effort.

This tendency persists despite the recent emphasis in the Canadian Forces (CF) on the need for arts and humanities in officer education. Between 1994 and 2001, following the atrocity of the Somalia Affair and the exposure of Canadian Airborne Regiment (CAR) initiation rituals, the Canadian government and the CF engaged in extensive inquiries into the sources of the moral breakdown within the CAR. They concluded that the problem was not restricted to the isolated case of the Somalia peacekeeping mission, nor to one regiment; rather, it concerned a

hotel rooms to eat chicken and watch porn. These parties have been *officially* banned by the college, but they are still, at the very least, part of the active cadet memory.

4 These three terms (homosociality, homoeroticism, and homosexuality) are not interchangeable. Homosociality is not generally associated with sex; it refers to social interactions that are exclusive to single genders. Sports, for example, tend to be homosocial, as do baby showers. Homoeroticism does imply the sexual, but, while it operates on the boundary or threshold of homosexuality, it does not necessarily include any kind of consummated same-sex desire. Only homosexuality unambiguously refers to same-sex sexuality.

fundamental failure of leadership and ethics and that the root of the problem lay in education (Haycock 2001: 6; Cowan 2001: 7). The Canadian military that grew through the Cold War had lost touch "with an increasingly vibrant, educated, and sophisticated society … [and the] system of military justice was out of line with an evolving rights-based Canadian criminal justice system" (Bercuson 2009: 34). In the early 1990s, only about half of the officers in the Canadian military had university degrees of any kind (Haycock 2001: 7), and this deficiency was deemed a significant reason behind the CF's "anti-intellectual conservatism, its rejection of reform, its failure to engage in challenging thinking" (Bercuson 2009: 34). So the disbandment of the CAR was only a small (though painful) part of the changes to the CF in the 1990s. By 1997, every officer in the CF was required to have a university degree that included "a compulsory and significant dose of arts, humanities, and social sciences education for all officer cadets, including those studying science and engineering" (37). At RMC, all students are required to take a first year introductory literature course, and all arts students must take an additional second-year course in twentieth-century literature.

With this requirement, CF policy implicitly aligned itself with the belief in the humanizing and liberalizing effects of university education. The CF has officially supported the constructive *use* of literature in the practical activity of making war. This alignment and support raises questions for literary scholars concerning the relationship between theory and praxis. Is one in danger of reducing the complexity and ambiguity of literature and literary theory to sensitivity training for the CF? On the other hand, perhaps literary theory's goal is its own adaptation to produce meaningful and lived human change. While I am not sure I can answer these questions, I will illustrate how my experience has moved through them in the context of sexuality in RMC culture. This chapter will outline how cadets react to and make use of literature in ways that allow them to interrogate and assess the homosocial, homoerotic, and homophobic culture of the greater and immediate worlds around them.[5] The first section deals with the homophobic response to literature in general; the second section discusses the homosocial, homoerotic, and homophobic nature of cadet life as revealed through cadet responses to James Hanley's *The German Prisoner*; and the third section illustrates the relationship between intimate bodily functions and male bonding through cadet responses to Hanif Kureishi's *My Beautiful Laundrette* and Erich Maria Remarque's *All Quiet on the Western Front*.

5 Many cadets are not at all homophobic, but there is a vocal homophobia at RMC, which indicates that a culture that unofficially promotes and sanctions homophobia still persists at RMC.

Poetry is Gay

Teaching in a military college, one is aware of the ambient homophobia circulating in one's classrooms. While hearing the casual homophobia is very hurtful, and while I am aware of the effect that it must have, consciously or unconsciously, on gay and lesbian cadets, I generally ignore it. Faced with the challenge of encouraging a group of mostly male cadets to seriously engage poetry, I initially choose to "pick my battles." Adopting an indifferent posture, however, became increasingly difficult, not because of my personal feelings of offence, but because I felt that I was not doing my job. I was teaching out of the stereotypical fear that my students would think that poetry is gay, and by failing to address that misconception, I was not only failing the profession, but failing the cadets, and, by extension, failing the liberal democracy that they serve.

We know that the arts have long been associated with the feminine and the queer. Alan Sinfield explains that, in the nineteenth century, creativity and the arts were coded along gendered and class lines when the wealthy leisure class came under attack by their middle-class detractors. The perceived idleness of the leisure class was feminized in comparison to the earnest, industrious and manly middle-class. In response, certain leisure representatives adopted the figure of the "dandy" in defiance of utilitarian middle-class hegemony. The strategy was "to repudiate manly, middle-class authority by displaying conspicuous idleness, moral skepticism, and effeminacy" (Sinfield 1994: 69). The dandy's same-sex association emerged "at the moment when the leisured, effeminate, aesthetic dandy was discovered in same-sex practices," which were dramatically instantiated and publicly distributed through the trials of Oscar Wilde (121). To this day, the artistic, the domestic, and the excessive are prominent associations with male homosexuality, reinforced in countless home-decorating and self-improvement television shows, and it is fair to say that many cadets have internalized these notions by the time they arrive at the college.

One of my most vivid encounters with the homophobic response to literature came in an introductory course that exposes students to a variety of genres from a variety of literary periods and cultural contexts. On the day in question, we were in a very canonical moment discussing Sir Henry Howard, Earl of Surrey's courtly love sonnet "Love, that doth liveth and reigneth in my thought," the last quatrain of which reads,

> And coward Love then to the heart apace
> Taketh his flight; whereas he lurks, and plains
> His purpose lost, and dare not shew his face.
> For my Lord's guilt thus faultless bide I pains. (2006: 354, ll. 9-12)

When I asked the students to describe the speaker of the poem, the first response came from one of the more lively contributors to class discussion, who said with calm disdain, "He's a fag, and if I saw the guy I'd kick his ass." Not only is Surrey straight, and not only is he an accomplished soldier, but this poem uses very martial metaphors to describe how love, "Clad in the arms wherein with me he fought,/Oft in my face … doth his banner rest" (Howard 2006: 354, ll. 3-4). So I was initially taken aback by this response to a poem that seemed, in its context, quite well aligned with a military ethos. After a few seconds of silence in which I adjusted to the blunt gay-bashing hatred of this characterization, I asked the student to explain why he thought the speaker was "a fag." Without missing a beat, he said, "He's like a chick." Clearly, more interrogation was required. "I see," I said, with mounting trepidation, "and what makes him 'like a chick?'" Again, no hesitation: "He's all emotional and weak." Click, click, click; it all falls so naturally into place, proving in this student's mind an indisputable equation: love and emotion are girly, and dudes who talk about love and emotion are fags; therefore, poetry is gay. In his way, this student was correct. In our homophobic and sexist Western culture we typically associate emotional disorder with femininity and weakness. It is enough to say, therefore, that this male speaker feels disempowered by his lack of control over his emotions. However, engaging this line of thinking in the detail that it deserves takes us into a discussion of the history, politics and theory of gender, which I was not willing or able to engage in at this stage of the class. I doubt very much that there was even time to achieve a measured response to the ideological underpinnings of the student's comments, which I found myself trying to unpack on the spot. In one instant, I could see why mandatory arts education was a difficult but invaluable component in the formation of the Canadian officer.

By ignoring homophobia to deflect from this association of poetry with homosexuality, I run the risk of leaving this association un-interrogated. Bruce Fleming's experiences at Annapolis reflect a similar failing to my own. Fleming tells us that when he introduced Wilfred Owen, he veered away from Owen's sexuality because, for him, some topics "are simply not worth the time and trouble—and … would simply be traps, black holes into which the class's energy would pour without ever coming back" (2005: 130). Perhaps unintentionally, Fleming uses a very homoerotic image of pouring one's energy into black holes, which demonstrates the always loaded silence of repressed sexuality, a repression that only deepened when the students asked about Owen's sexuality on their own initiative. For instance, one student could only come to terms with Owen's sexuality by realigning Owen's masculinity through the intimacy of violence: "I always thought that fags wouldn't kill anybody. I mean, I thought they were sort of fruitcakes. But Owen, he killed a lot of men. And he got a medal. It changed my views, reading that. I mean, he was ok" (131). As Fleming admits, this is a disturbing response. In its own way, it is as disturbing as the violence *against* gay men in the response to Surrey's speaker. In the masculine military culture of Annapolis, the midshipman can only conceive of Owen's passionate and tender longing *for* other men in terms of the violence Owen committed *against* other men. In the case of both Surrey and Owen, both soldiers who trouble the definition of manliness, violence is the route through which the security of the homosocial military culture of cadet life is re-asserted within the disruptive discursive space of the classroom.

Getting Cocked

In an effort to engage with this troubling tendency in the military academy classroom, I decided to teach James Hanley's 1930 story *The German Prisoner*. This story depicts two ugly, animalized and disempowered working-class World War I soldiers named Elston and O'Garra, from the slums of Manchester and Dublin respectively, who get lost in a surreal blasted warscape of mud and fog. As they grow more frightened and more dependent upon one another, the soldiers come upon a young and beautiful German soldier whom they torture, rape, and murder. The story's representation of torture in the military context obviously resonates with the Somalia Affair and the atrocities at Abu Ghraib, but the story involves a complex of other themes, some of which pertain to the current discussion of the relation between the aesthetic and homoerotic violence.

To begin with, in stark contrast to Elston and O'Garra's very bodily, even scatological, reality, the German soldier is clearly an aesthetic object:

> He was a youth, about eighteen years of age, tall, with a form as graceful as a young sapling, in spite of the ill-fitting uniform and unkempt appearance. His hair, which stuck out in great tufts from beneath his forage cap, was as fair as ripe corn. He had blue eyes and finely moulded features. (J. Hanley 2006: 25)[6]

With his "finely moulded features," the German prisoner stands out as an aesthetic idealization in a story that is otherwise basely realistic. In contrast to the debased realism of the rest of the story, this figure is as out of place in this landscape as his uniform is ill-fitting. His youth is literal as well as figurative, recalling a lost innocence that Elston and O'Garra cannot share (Rice 2002: 76).[7] He is also rural and natural, "graceful as a sapling" and "fair as ripe corn," where Elston and O'Garra are urban and corrupt. Recognizing the German soldier's distance from their own experience, Elston and O'Garra refer to him in scatological and homophobic language in order to reduce him to their level of experience: they refer to him as a "sod," a "bugger," and curse him saying, "Shit on you" (J. Hanley 2006: 25). Their speech quickly turns to action, as O'Garra urinates on the beaten prisoner, and the two men subsequently assault him in a scene of deeply homoerotic violence:

> The two men fell upon the prisoner, and with peculiar movements of the hands began to mangle the body. They worried it like mad dogs. The fog had brought

6 Hanley's story was originally published as a limited edition by the anarchist bookseller Charles Lahr. In 1935, it was banned for obscenity, along with his novel *Boy*.

7 Following Rice, I often teach this story in terms of its difference from the representations of officer-class poets like Owen and Sassoon. Many working-class soldiers, like Elston and O'Garra, came from a preexisting state of economic victimization and so did not identify with the prewar golden age of Edwardian innocence.

> about a nearness that was now driving them to distraction. Elston, on making contact with the youth's soft skin, became almost demented. The velvety touch of the flesh infuriated him. Perhaps it was because nature had hewn him differently. Had denied him the young German's grace of body, the fair hair, the fine clear eyes that seemed to reflect all the beauty and rhythm and music of the Rhine. Maddened him. O'Garra shouted out:
> "PULL his bloody trousers down."
> With a wild movement Elston tore down the prisoner's trousers.
> In complete silence O'Garra pulled out his bayonet and stuck it up the youth's anus. The German screamed.
> Elston laughed and said: "I'd like to back-scuttle the bugger."
> "Go ahead," shouted O'Garra.
> "I tell you what," said Elston. "Let's stick this horse-hair up his penis."
> So they stuck the horse-hair up his penis. Both laughed shrilly.
> A strange silence followed. (J. Hanley 2006: 35)

Once again a thing of beauty, not only with "finely moulded features," but now reflecting all the "beauty and rhythm and music" of the Rhine, is put in its place through homophobic violence. The violent feminization of the German prisoner is an attempt to reassure the soldiers in their shared masculine power, but it clearly exposes the insurgent homoeroticism upon which the worship of that masculine power is based. For example, Elston's dementia mounts as "nearness" becomes "contact with the youth's soft skin" and a "touch" of the velvety flesh. Further, when Elston states his desire to "back-scuttle the bugger," he names the prisoner for the act he wishes to commit. "Nearness" is the key word in the description of this scene: the nearness of Elston and O'Garra, the nearness of male bodies, the nearness of revulsion and desire. Elston and O'Garra both embrace and banish that nearness. As they draw ever-closer to each other, they destroy an object of desire that threatens to expose their too intimate nearness. Notice how much of this scene revolves around exchanges between Elston and O'Garra: O'Garra's command leads to Elston's compliance; Elston's expression of desire leads to O'Garra's permission; Elston's suggestion leads to joint action. This is an intimate experience shared in a give-and-take between the two men that establishes a nearness that is disgusted by its own desire and which banishes that desire in the process of indulging in it. This dynamic is remarkably similar to Butler's and Burke's discussions of the homoerotic/homosocial nature of military male-bonding.

In *The German Prisoner*, therefore, we have a war story and a work of art that depicts the violent homoerotic destruction of an aestheticized youth, beauty, and innocence. When I first led the discussion of this story, I was cautious about exploring the issues that I've just outlined, but I needn't have worried, for this excessively brutal scene captured the imaginations of my first-year cadets who had recently gone through an initiation in which their own youth and innocence had been ritualistically replaced by masculine power. The discussion began when one cadet pointed out that the homosexual rape figured in *The German Prisoner* is performed in speech in cadet life. Specifically, he referred to "getting cock" (Burke 2004: 121).[8] "Getting cock" denotes the experience of subjection to one's superiors, particularly when being humiliated or punished; it is therefore an expression that applies to much of the first-year cadet experience the world over in which cadets go through the same ritual abasement as "three-fourths of the student body assumes the role of the master and one fourth reluctantly plays the part of slave" (19).

On their own initiative, my students explained in their own terms that the brutal power matrices in *The German Prisoner* are replicated in the experiences that "getting cock" applies to. The expression "getting cock" performs the homoerotic as it homophobically denigrates homosexuality through the violence of rape. Their response is supported in greater detail in John James Ford's 2005 novel *Bonk on the Head*, which is based on his experiences as a cadet at RMC in the early 1990s. A note of caution is called for when using Ford's novel as a primary source. Ford's depiction of his life at RMC is not a straightforwardly objective retelling of his experiences, but, as he explained in a public reading at RMC in 2007, a fictional amalgamation of events that happened to himself and to various other friends. Also, his experiences predate the recent changes to CF policy on education. On the whole, however, one may easily argue that it is a true account of certain abiding cultural trends at RMC.

Ford uses the term "getting cock" throughout the novel as he explicitly recounts the homophobic forms that cadet humiliation can take. For instance, the protagonist, Herbert Kempt, is forced to adopt an anal-sexual position with his roommate while the rest of the squadron watches so that, as the fourth-year cadet explains, homosexuality can be defended against: "You two stay like that! We've got to know who's queer, right, so the lads can watch their wee-wees in the shower" (Ford 2005: 152).[9] Herbert resents that for the next few weeks he is known by various homophobic names, such as "Recruit Cockchugger, Ballsucker, Fudgepacker, Dicklicker, Assmuncher," but once he passes the stage of cadet initiation, he adeptly deploys homoerotic language with his peers (153). Walking behind one friend, he adopts an aggressive sexual posture, saying "Hey, little man … Nice ass," to which his friend replies, "Thanks, buckshot … You want to pump it full of cream?" (Ford 2005: 222).

This intersection of the homoerotic and the homophobic in *The German Prisoner* and in cadet life at RMC conforms to Eve Kosofsky Sedgwick's definition of homosociality as existing along a continuum with homosexuality.

8 Burke cites Moore 1993 to emphasize the ubiquity of anal sex in cadet language: "Cadets litter their conversation with references to homosexuality; Moore lists 70 slang terms for anal sex alone! Someone who is gay is a 'freckle puncher,' 'poo pirate,' 'turd burglar,' 'bung-hole buster,' or 'bum puncher.' To be 'goat fucked' is to be required to perform an unpleasant task" (Burke 2004: 121).

9 For more examples of homoerotic initiation in military and cadet culture, see Burke 2004: 15, 44, and 60-77.

Part of homophobia's cultural power comes from the "similarities between the most sanctioned forms of male-homosocial bonding, and the most reprobated expressions of male homosexual sociality … For a man to be a man's man is separated only by an invisible, carefully blurred, always-already-crossed line from being 'interested in men'" (Sedgwick 1985: 89). Homophobia is about clearly drawing a line between homosexuality and intense homosocial male bonding. It is part of what Butler describes as the "heterosexual logic that requires that identification and desire be mutually exclusive [which] is one of the most reductive of heterosexism's psychological instruments: if one identifies *as* a given gender, one must desire a different gender," as if there were such a thing as two genders and not a range of ways to "be masculine" and "be feminine" (1993: 239). According to Fleming, blurring the lines between identification and desire is also central to the culture of Annapolis:

> Thus male bonding is still at the center of the Academy. Our male students know it, but, in our attenuated gender climate, are not allowed to say it … Male bonding is socialization not only in the absence of the Female (if not in the absence of women), but also to a certain extent the rejection of the Female (which tends to mean the rejection of women). It implies not only the denigration of the Female, but the worship of the Male. And this means that it is profoundly homoerotic. The paradox of the Academy is, therefore, that it is both homoerotic and homophobic. (2005: 147)

In the military worship of the male, identification and desire meet. In *Bonk on the Head*, Herbert recalls a blond cadet sauntering into his room at the beginning of summer training: "He was dressed in PT gear that fit him perfectly, and though he was a touch shorter than either of us, his body was imposing … He had the kind of body Dad would have called a 'brick shithouse.' It seemed impervious to pain … Soaked with sweat, he extended a meaty palm. Attached to his hand was an arm thickly braided with muscle" (Ford 2005: 260). After brief introductions, the blond cadet strips down, shows Herbert and his friend his "neatly shaved bag," and invites them to touch it (261). There is a clear worshipfulness in Herbert's description of the blond cadet, and the *desire-to-be* the cadet is emphasized by his thoughts of his father, whose approval Herbert seeks throughout the novel with self-mortifying desperation. Herbert touches the blond cadet's shaved scrotum, and, on contact with the soft skin of the velvety flesh (to borrow words from *The German Prisoner*), the line between identification and desire, or *desire-to-be* and the *desire-for*, breaks down.[10]

As Fleming indicates, the breakdown of the identification-with and desire-for maleness inevitably involves the rejection of the female. Female cadets make

10 This scene in *Bonk on the Head* approximates a custom in RMC cadet culture called the "shaving party," in which two or more cadets gather in a group to shave their scrota.

up barely 20 percent of the student population at RMC, and their voices in class are subdued by more than mere numbers. Several female students have privately expressed reluctance to voice opinions that may be construed as feminist, for fear of alienating themselves from their male peers. Similarly, Nancy Taber, a cadet at RMC during the 1990s, argues that women at RMC are forced to become "social males," which includes adopting a violently sexist stance against women (Taber 2005: 293).[11] Taber's most revealing example of the rejection of the feminine is a chant that is ingrained into every cadet, part of which runs: "Shrapnel, Cordite, NCT, R—M—C! Hooah!" (293). Although Taber explains that she never saw the meaning of the acronym NCT written anywhere, it is universally understood to mean "No Cunt Tonight," a fact that current RMC cadets confirm.[12] In this chant, which is heard in the masculine arena of competitive sports, female voices are overpowered by the majority male voices, and female bodies are performatively excluded. Burke records the same ritualized banishment of the feminine in many military songs and chants, like the following:

> Now Suzie said, "It's me or the corps,
> I can't take this life no more."
> So I looked at her with a big ole grin.
> I haven't seen Suzie since I don't know when. (Burke 2004: 29)

Such songs and chants "devalue all nonprofessional affiliations and mark a soldier's passage from civilian life to combat by encouraging masturbatory compensation: 'I don't want no teenaged queen./I just want my M-16'" (30). The worship of the phallic M-16 and the fetishization of the penetrating bayonet[13] in *The German Prisoner* demonstrate that, in cadet culture, "getting cock" refers to a complex social and culture negotiation of masculine identity in the military context, and it is not always a bad thing. I now regularly use the term when teaching *The German Prisoner*, and students always explain its subtleties with a distinctly reverent amusement. They may all hate "getting cock," but at least it's *their* cock. As Herbert's friend in *Bonk on the Head* puts it after the blond cadet has gone for his shower: "Shit … I'm fuckin' more than ready for the cock these fuckin' cunts are gonna dish out" (Ford 2005: 262).

---

11 I am indebted to Naval Cadet Robyn Brink for directing me to this article.

12 As one male cadet put it in an email, "As to NCT it has been said to mean that [No Cunt Tonight], as well as No Cock Tonight (referring to harassment [by superiors]) and possibly other things. But knowledge of the origins of this cheer is very scarce since it has apparently been around longer than anyone who is still alive remembers." A female cadet had heard another permutation: "No Cunts in Training." Another female cadet admitted that even after four years at RMC, she did not know what NCT stood for.

13 Hodges 2008 demonstrates links between masculinity, bayonet fetishization, atrocity, and prisoner killing that provide historical insights to Hanley's story.

Epistemology of the Thunder-Box

The homoerotic homophobia at RMC generally confirms Lee Edelman's contention that "in the dominant homophobic culture that still persists in the world, homosexuality primarily comes into focus only as the undoing of one man's authority by another" (Edelman 1994: 54). Therefore, the intimacy between men in military cultures, where different levels of authority are taken for granted, is deeply implicated in power, especially where the line between identification and desire is most blurred. In military training, the site at which this power is most often exercised is the cadet's body. The cadet's physical movements, clothing, bodily functions, sleep, diet, and physical form all fall under the authority and management of the CF. This third section of the chapter, therefore, focuses on that critical site of desire and power: the body. I approach this topic through a discussion of the communal or public washroom as a site at which the public and private control of the body and bodily functions is exercised with a particularly heightened awareness. In the classroom, these issues arose in the context of discussions of Hanif Kureishi's 1985 film *My Beautiful Laundrette* and Erich Maria Remarque's *All Quiet on the Western Front*. The first text exposes students to hybridized postmodern identities as these identities negotiate forms of public power and experiences of private desire as symbolized by the laundrette, a space with strong psychic connections with the public washroom. The overt homosexuality that some cadets objected to in *My Beautiful Laundrette* is then contrasted to the more sanctioned forms of male-bonding in *All Quiet on the Western Front*, which cadets, once again, aligned with the cultural practices of cadet life at RMC.

My Beautiful Laundrette is set in Thatcherite London suburbs and revolves around the romantic relationship between two young men, Omar and Johnny. Omar is an upwardly mobile Pakistani businessman, and Johnny is a white working-class former fascist. Many students do not read the screenplay before the film-viewing, and those who do read it in advance remain quiet about the gay relationship at its heart, perhaps as interested as I always am in the spectacle of its revelation. The 50 minute class is perfectly timed so that the first class ends immediately after the first screen kiss. The little moments of intimacy leading up to this kiss sometimes elicit homophobic comments and jokes, but are generally not read as clear signs of homosexuality. The kiss itself can elicit roars of outraged disgust. Part of the film's success is in its treatment of the gay relationship. The two central characters who are so clearly marked in terms of class and race by clothing, speech, and skin color are almost completely unreadable sexually, and the film uses no word that names its queerness.[14] The students therefore operate within the presumptive heterosexuality of normative culture until it is visually shattered before their eyes. Even then, the ensuing discussion invariably questions whether or not Omar and Johnny are in fact gay. Some argue that their sexual behavior is an aberration of

14 The only references to either character in a queer context are several offhand uses of the word "bugger."

their normally heterosexual identities; others argue that they must be bisexual, pointing to Omar's flirtation with his cousin Tania. The film's refusal to inscribe these dissident male bodies in fixed visual and spoken terms invites this kind of uncertainty and illustrates the anxiety that leads to the heteronormative need to put sexualities in their places, especially within institutions like the military that are so emphatically gendered.

One might even argue that Kureishi's unspoken representation of queer bodies is the inverse of the sexual culture of RMC, which is so voluble about queer sexuality. In British and North American culture, heterosexuality is so normative that it does not bear thinking, and most people are presumed straight until they are revealed as otherwise. But intimate male-bonding in the military does not allow the same presumption. In normative British and North American culture, once someone is revealed to be gay/lesbian/bisexual/transsexual, their private lives become public by virtue of their difference. In military culture, the private lives of soldiers and cadets are made extravagantly public to preserve heterosexual sameness.

Kureishi directly engages this public/private negotiation by choosing a laundrette as the primary site of action. When I ask the students why he does this, I quote a number of passages. Omar proclaims that the "whole stinking area's on its knees for clean clothes" (Kureishi 2006: 2967). His cousin Tania believes that Omar's father doesn't want his son working in the "shitty laundrette" (2953). As Omar's father puts it, he wants his son "out of this underpants cleaning condition" (2974). But Omar stubbornly holds onto his vision. Just before his first kiss with Johnny, he explains his plans: "I've had a vision. Of how this place could be. Why do people hate laundrettes? Because they are like toilets. This could be a Ritz among laundrettes" (2963). Laundrettes are like toilets (in particular, public washrooms), because they are sites at which the public and the private converge, where underclothing is held up in the public light of day, and where cleanliness shows its precondition of filth, where bodies, in all of their messy operations, are revealed. This site of transgression is the ideal location for a film dealing in so much racial, social, and sexual border-crossing and potential "contamination" of the condoned narrative of 1980s Britishness.

In his essay "Tearooms and Sympathy or, The Epistemology of the Water Closet" Lee Edelman discusses the "intertwining of sexuality and nationalistic ideology in the United States" (1994: 148). Specifically, he examines the public figurations of homosexuality in 1950s and 1960s US and "posits homosexuality as a legible phenomenon while simultaneously acknowledging the frequency with which it manages to escape detection" (154). His argument centers on the paranoia around, and official surveillance of, "the institutional men's room [which] constitutes a site at which the zones of public and private cross with a distinctive psychic charge" (158). Quoting a 1964 article in *Life* magazine which celebrates the American public washroom, Edelman discusses these spaces in terms that strongly recall Omar's vision of a "Ritz among laundrettes":

> "Bedecked and bejewelled and splashed with color, the bathroom is blossoming with a flair unapproached since the fall of the Roman Empire." Thus caught between its honorific associations with industrial progress and hygienic purity, and its more pejorative associations with weakness, luxury, and a decadent indulgence of the perverse, the American bathroom in 1964 constituted an unacknowledged ideological battleground in the endless—because endlessly anxious—campaign to shore up "masculinity" by policing the borders at which inside and outside, same and different, self and other are the psychic stakes at risk. (159)

These ostentatiously public spaces must be excessively beautiful and clean to divert attention from the private bodily functions that are secreted away; however, like military culture, they offer an ostentatious announcement of the unspoken. The washroom's intimate and disavowed proximity to sexual organs, bodily functions, and the opening of the male body, reminds us, Edelman argues, of Julia Kristeva's words: "It is thus not lack of cleanliness or health that causes abjection, but what disturbs identity, system, order. What does not respect borders, positions, rules" (162).

Ford's depiction of Recruit Term (now called First Year Orientation), where the borders of masculinity are ritually inscribed upon each cadet, is closely bound to the control and violation of bodily function. For example, after thirteen hours of being denied the time to urinate, Herbert "fantasized about the urinals: fragrant piss pucks, cold metal taps, tile, grout, porcelain and condensation, piss, piss, flush" (Ford 2005: 118). When he finally steals a moment to urinate, he does so in his dorm sink, and the experience is a sensual one: "I unzipped and was almost pissing before I could even aim at the sink. Probably still the best piss I've ever had—the kind where you get shivers, a cold knife singing right up your spine" (119). As he indulges in this almost orgasmic release, his roommate "just gape[s], staring at [Herbert's] penis with his mouth open" (119). When the roommate, too, finally urinates in the sink, he experiences shame and satisfaction: "Mack stood there for a moment, and then, slowly, undid his fly. He did it sadly, as if by pissing in this here sink he was letting down his mom back in North Bay and everything she stood for. Holding onto the towel rack he broke the seal and let out a low moan" (122). For thirteen hours, the cadet's body has been strictly controlled. At the moment when that control is momentarily lifted, a series of borders are crossed as shared male sensuality located in uncontrolled bodies operates private bodily functions in public spaces which undermine, "let down" or "break the seal" on a pact with normative culture and with their former civilian selves, opening up new relations between men.

Ford's depiction of cadet washroom culture and of the control and opening of bodies extends to more explicitly fluid relations between men. On "Dognight," the night before the obstacle course that marks the end of a month-long orientation/initiation process, a group of third years are "bestowed the privilege of roughing up the fresh meat" (194). After being taken on a midnight jog, forced to do sit-ups and wind sprints, pelted with eggs, smeared with food and cosmetics, and taunted with a hot iron, Herbert has wet himself and sobs uncontrollably as the cadets are marched into the washroom:

> We all lay face-down and they turned the lights off. The floor tiles stank. My face was on fire. People were moaning and shivering up against each other. I think I might have been moaning, too.
>
> I felt a warm trickle hit my raincoat, my neck, my toque, my face. "They're fucking pissing on us, man," Billy whispered to Josh. I was hoping he would laugh. But he didn't. He didn't make a fucking sound. (Ford 2005: 196-197)

Here again, borders of public and private, inside and outside, self and other, are crossed in the charged space of the communal bathroom through the opening of the male body and the transmission of bodily fluids. Behind the ordered masculinity of the uniform lies a fluidity that disturbs identity, system, order, borders, positions, and rules. This fluidity and transgression, however, is the paradoxical other that sustains the public order. Indeed, as Ford presents it, the obstacle course at the end of Recruit Term perpetuates this spectacle of disavowed scatological humiliation: "We were swimming through a sewer, and all these people, even our parents—especially our parents, were laughing and clapping and spectating" (199). This scene is the ritualized military analogue to the "Ritz among Laundrettes," and the "bedecked and bejewelled" public washroom.

Although many cadets were hostile to, evaded, or flat out denied the homosexuality in *My Beautiful Laundrette*, discussions of the social, psychic, and symbolic nature of washrooms in the context of homosociality and homosexuality led to discussions of the washroom as a communal space in RMC cadet life. The practice most directly relevant to our current discussion is the "log party," a custom which crosses similar borders to those described by Ford, but without any of the humiliating trauma. "Log parties" occur when a group of male cadets occupy a row of washroom stalls and engage in communal defecation and communication. A similar custom is the "shaving party," in which several male cadets gather to shave their scrota in each others' company. These are two very bodily forms of male bonding associated with ostensibly private acts involving genitalia and anality.

Given this physical homosocial intimacy, one should not be surprised at how easily cadets identified with an early scene in Erich Maria Remarque's *All Quiet on the Western Front*, in which the principal characters gather to defecate in a circle of thunder-boxes, or empty ammunition boxes. As the novel's narrator, Paul Baumer, explains,

> Being forced to do everything in public means that as far as we are concerned, the natural innocence of the business has returned. In fact, it goes further than that. It has become so natural for us that the convivial performance of this particular activity is as highly valued as, well, holding a cast-iron certainty of a hand when we are playing cards. It is not for nothing that the phrase "latrine humour" has come to mean all kinds of gossip; these places are the army equivalent of the street corner or a favourite bar. (Remarque 1996: 6)

Here the private act is unquestionably translated into a public one, and zones of privacy become zones of community. Several male cadets readily recognized this comradely communion over bodily functions and identified it with their "log parties." Even Herbert in *Bonk on the Head* comes to a similar conclusion near the end of the novel:

> JP tells us to feel his balls and we do. There's nothing weird in it. The rugby boys piss on each other in the shower. We stake claims in the sexual, the oral, the scatological. Since Dad died, I see how our minds are on hold, our thoughts de-individualized; only our bodies belong to us, and then so infrequently there is no room for social convention. We must stake our claims between the runs, the drill, the push-ups. We stake our claims in our bodies, and feel comforted, grounded, brotherly. (Ford 2005: 261)

Of course, these homosocial rituals are never confused with the homoerotic and the homosexual. While students quickly identified with the anal-centered communion on the thunder-boxes, they evaded discussion of Paul's intimate and homoerotic attachment to his older mentor-friend, Kat. Even passages like the following are explained in the terms of the uniform masculine comradeliness exercised in the "log party":

> We don't talk much, but we have a greater and more gentle consideration for each other than I should think even lovers do. We are two human beings, two tiny sparks of life; outside there is just the night, and all around us, death. We are sitting right at the edge of all that, in danger but secure, goose fat runs over our fingers, our hearts are close to one another, and time and place merge into one—the brightnesses and shadows of our emotions come and go in the flickering light of a gentle fire. What does he know about me? What do I know about him? Before the war we wouldn't have had a single thought in common—and now we are here, sitting with a goose roasting in front of us, aware of our existence and so close to each other that we can't even talk about it. (Remarque 1996: 66)

Furthermore, this love is not purely platonic, but expresses itself through the love of the body: "Half asleep, I watch Kat as he lifts and lowers the basting spoon, I love him; his shoulders, his angular, slightly stooped frame—and then I see the stars and woods behind him, and a kindly voice says words to me that bring me peace" (66). There is some magical *cordon-sanitaire* that lets us read these passages filled with traditional romantic tropes and stop just short of seeing any homosexual love. In Paul's love for Kat, which adopts the language of sameness and equivalency, we have a strong contrast to the more obvious and violent homoeroticism of *The German Prisoner*, which is based in difference and power, but we still police the boundaries of that desire as stubbornly as Elston and O'Garra do. We must not *ask* after the nature of Paul's love for Kat; we must not *tell* that there is anything more than the comradely affection between men in peril. We must not threaten the sanctity of military manhood.

Conclusion

When students read and discuss literary texts, they not only discuss the Renaissance sonnet, or the nature of 1980s British national identity, or the experience of World War I combat; they also face some of the central ethical questions of becoming officers in the CF. The Canadian military has emerged from a period in which its values were out of step with the increasingly rights-oriented value system of an increasingly educated Canadian civil society, and the post-Somalia Affair CF is now committed to the liberalizing effects of humanist education. Therefore, the private rituals of RMC in particular and the CF in general must be examined in the context of public discourses of identity, masculinity, and Canadian military culture. The challenge of overcoming the classist, sexist, and homophobic association of literature with triviality, femininity, and homosexuality (or even the generally odd, freakish, "liberal", hippie, and deviant) is one of literature's greatest use-values to the CF. Obviously, a military culture that, in accordance with the Charter of Rights and Freedoms, welcomes women and LGBT members will find its rituals and traditions adapting to less limited forms of gender and sexual identity, and the unruly and transgressive diversity of literature will help this ongoing process.

Blurring the boundaries of identification is, I argue, one of the most important elements of officer education, which must cultivate critical thinkers who are capable of representing the full range of Canadian citizenship and communicating with the diversity of global communities (and the diverse genders and sexualities of those communities). There is still some work to be done before the CF fully accommodates its LGBT members and before it serves Canadian LGBT citizens, and that work begins, as this chapter has tried to demonstrate, by recognizing that heterosexual masculinity is *already and inescapably queer* (not to be mistaken with homosexual). Canadian military culture would do well to understand its own queerness, its own dependence upon a broad range of relations between men. When it can do so, it will more truly fulfill its commitment to an ever-maturing CF.

PART III
Teaching in Professional Military Schools

Chapter 9

Navel Gazing Google Deep: The Expertise Gap in the Academic-Military Relationship

Thomas Bruscino

The United States Department of Defense employs professional academics in essential advising, teaching, researching, and writing positions throughout the armed forces. Despite the prevalence of such individuals, there remain significant misunderstandings between service members and academic civilians. Many of these issues revolve around clashes in two very different cultures, and leaving aside the more obvious issues of jargon, politics, work schedules, and command structures, the most pronounced cultural difference is in the area of expertise. Academics and most members of the military have very different understandings of the meaning of the word, and those differences lie at the root of a significant gap in the academic-military relationship.

Background

This observation stems primarily from personal experience. I do not come from a military background, and I have never worn the uniform. What is more, like most Americans I spent most of my life isolated from the military. My grandfathers did their part in World War II, both in the Navy, but that was a different time. My dad got a draft card for Vietnam, but not until the war had wound down to almost nothing, so he never really had a chance of going. I grew up in the suburbs of Cleveland, Ohio and Denver, Colorado, played sports, went to school, and never seriously thought about serving in the military.

In all that time, I had little or no contact with the military. My small college did not have an ROTC component, and the ROTC folks at my graduate school generally kept their heads down. I became a fulltime academic graduate student, planning on working in an academic environment. Despite my lack of contact with the military, war, and what would compel people to fight, always interested me, so I did study military history, specifically American military history, and gained an understanding of historical issues of civil-military relations. But even that experience came from a decidedly academic point of view. I was at a civilian university and intended to teach civilian students at civilian schools.

Yet in the time since receiving a doctorate in United States military history, I have worked as a historian at the US Army Center of Military History at Fort

McNair in Washington D.C., a historian at the US Army Combat Studies Institute at Fort Leavenworth, Kansas, and as an assistant professor at the Army's School of Advanced Military Studies in the Command and General Staff College at Leavenworth. It would be an overstatement to call my interaction as an academic with military culture a shock, in large part because the organizations I worked for already employed civilian PhD historians and were about as far away from the sharp end of the military as they could be. But my background as a civilian and civilian academic has engendered a sensitivity to the divergence between that life and my work experiences, so what follows will initially focus on the differences.

The Center of Military History has a variety of missions, including the writing of official history, tracking unit lineages, gathering historical data for later use, and running Army museums. The Center also provides more direct historical support, and I worked in that branch. Generally, we wrote information papers in response to history-related questions—for example, on average how long have troops been deployed for disaster relief to hurricanes in the past two decades?—from the Department of Defense, Congress, the public, and (mostly) the Army staff. We also wrote the yearly Department of the Army Historical Summaries, cobbled together without citations from reports from commands and headquarters across the force. Both of these tasks had great value, but neither required anything like the skills garnered in graduate school. In fact, both jobs involved a great deal of recycling of older information from previous requests. Borrowing from older versions or reusing old material is a fairly standard characteristic of military staff work, but any academic would immediately recognize how much such borrowing resembles the "p-word." Considering how many Army organizations and units have civilian command historians with the task of submitting to the center their yearly historical reports, the experience was not limited to the center itself ("Army" 2007). Nor does this sort of work seem to be limited to the Army. A recent critique of the US Naval History and Heritage Command noted that one of the stated capabilities of the command was that "Navy leaders in the Pentagon could call to get a custom report on historical precedent to aid in making a difficult decision" (qtd. in B. Hanley 2009: 10).

The Combat Studies Institute has its own mission, one with a presentist focus. The institute conducts staff rides, writes first-cut histories of current conflicts, gathers oral histories with veterans of contemporary conflicts, and produces historical studies in support of various Army concepts and missions. Again, all of these tasks have clear value to the contemporary Army. For my part, I worked mostly on writing the historical studies. Almost all of the topics for such studies are directed from above. From somewhere high in the chain of command, generally the office of the Combined Arms Center command, a list of topics of interest is provided; examples include counterinsurgency in Africa, military public affairs, the use of unmanned aerial vehicles, interagency concerns, and so on. The leadership of the Combat Studies Institute selects those they deem most appropriate and passes them on to the historians, along with deadlines usually in the 3-6 month range. Due to the nature of these assignments (that is, how is this topic relevant to the contemporary force?) and because of the truncated work time, the historians generally cannot do a great deal of research and have to pick some sort of very specific analytical lens through which to look at a series of historical case studies. It is a far cry from the much more open research agendas in the academic world. I am not familiar with the process of topic assignment for the Army's Strategic Studies Institute, the Air University's various series of publications, or the Naval War College's Newport Papers, but in general their products are similar to those produced by the Combat Studies Institute.

Even though the School of Advanced Military Studies is a select advanced graduate program, the standards and practices at the school are much more in line with the rest of the military education system than any academic graduate school. For the most part, the school has a standardized curriculum. In the pursuit of specific learning objectives, all of the students receive the exact same course material on paper. Regardless of the specific classroom or specific professor, every lesson, every reading within the lesson, and even discussion agendas for class instruction are exactly the same. In general, one professor writes the lesson, selects the readings, and lays out the discussion, and other professors teach from that material. Obviously, there are differences in style and emphasis within the classes based on the direction the students take, but the foundational material is almost always identical. Even when professors choose some of their own material, the standardization is far in excess of most undergraduate programs, let alone graduate courses.

Although most of this discussion comes from personal experience, the evidence indicates that the production of recycled information papers and rushed monographs and the teaching of unfamiliar topics in an over-centralized curriculum are common in the history and education realms of the military. This kind of work runs counter to so many of the intellectual underpinnings of advanced graduate education that the academic almost has to practice selective amnesia to get through the day, forgetting the professional standards, broad contextual knowledge, and deep research that are the hallmarks of gaining a terminal degree.

What is more, military culture seems to have little to no awareness that that is exactly what the PhDs in their midst are doing. The military's concern, quite fairly, is with the application of academic knowledge to their craft. If an academic cannot explain why a subject matters in the most practical of terms, that subject appears to have no value. Real expertise, and the less obvious benefits of such expertise, can too often get lost in translation. When academics introduce some of the contentious but highly specialized debates from their respective fields into this environment, they are often accused of arguing over how many angels can dance on the head of a pin, or, less charitably, for contemplating the fuzz in their navels.[1] Lest they be accused of too much navel gazing, hundreds of academics in the military's employ leave behind their professional standards and spend most of their time doing work that they know any reasonably bright master's student could handle.

1 For commentary on anti-intellectualism in the armed forces, see Matthews 2002a and 2002b.

Healthy Tension?

By necessity, this has been a one-sided account. There is something more going on, because if that was the entire case all of the very bright officers with master's degrees would not feel compelled to bring academic PhDs into their institutions. This is an old issue, long evident in the military's use of academics and history.[2] The military seems to know it needs something more to tell its history, provide the intellectual foundation for its actions, and to educate, rather than just train, its people. When members of the military offer opinions with weak or no evidence or provide analysis without much research, other soldiers will call them out for only being "Google Deep" or "Wikipedia Deep" on the subject, meaning the individual in question went to Google, did a search, and repeated the information found in the first link at the top of the page.

My three examples, despite the shortcomings noted above, are all indicative of the military's efforts to get past the superficial. The fact that there is a Center of Military History—along with all of the other academic-style institutes, centers, and think tanks across the Department of Defense—shows a serious concern for bringing academic inquiry into the military.[3] For years the center, with a great deal of support from military patrons, has resisted attempts to rush the production of official histories. They see their mandate as providing thorough, well-researched, well-written, even-handed, and academically impeccable accounts of the Army's history, even if the relevance of those histories is not immediately evident to the contemporary force. The official histories of the Army in World War II set this standard, and it has been carried on through Korea, Vietnam, and the many thematic studies published throughout the years. Indeed the academic field of military history would be greatly impoverished if not for the efforts of the Center of Military History.[4]

The Combat Studies Institute offers similar benefits. The military could have non-academics produce the quick turnaround studies of issues of immediate relevance to the Army, but instead gets individuals with advanced degrees to do the work. These academics try, given the time, to bring their education to the products. They succeed in some cases better than others. But there are two pretty important benefits to such work. First, when it comes to contemporary history, the fact that trained academic historians apply their skills to current issues means that they apply a different and more rigorous academic standard—including an accounting of relevant literature and thorough documentation of sources—to the writing of the first draft of history than the journalists who normally dominate the field.[5] Second, by being forced to approach history through issues of relevance

2 Two excellent overviews are Millett 1992 and Sinnreich 2006.

3 For an overall review see the roundtable discussion in Kohn 1997.

4 A recent debate on the center's activities played out in the center's periodical, *Army History*. See Fontenot 2008, Hammond 2008, and Stewart 2009.

5 See Wright and Reese 2008.

to the contemporary force, they potentially introduce new ways of understanding military history to the wider field. The focus on combined arms warfare or the use of private contractors across the range of American military history are good examples, but there are many more.[6] In its employment of academics, the Combat Studies Institute makes a contribution to both military and academic cultures.

The School of Advanced Military Studies is an even more direct example of the Army trying to go beyond what the military can find internally. As the authors of one review of professional military education wrote:

> US Army Training and Doctrine Command (TRADOC) launched SAMS as an optional course for volunteers who completed the standard staff college program. The aim was to attract the brightest from the core combat specialties to examine the present and future of their profession in an intense, graduate level course, equipped with an understanding of the process based on the first year of study at Fort Leavenworth. In effect, the creation of SAMS was a tacit admission that the regular intermediate level program was not serving the needs of commanders in the field. (Holder and Murray 1998: 84)

The Air Force and Marine Corps have created similar schools for themselves. Academic PhDs have played a key role in ensuring that the content of the program at those schools stays at a graduate level, even if there are ongoing debates about what constitutes graduate level learning in military arts and sciences.

The fact is that there is an ongoing tension within all of these organizations and across the Department of Defense between the clear need for direct relevance of educational material and an intuited need for the greater depth that only academics can provide. American fighting men and women are practitioners, and they have to get things done. If they cannot see how their education helps them do their jobs, or even worse, if they think it might hurt them (navel-gazing), then they do not have much use for it. But they sense that not all education has direct or clear benefits to their everyday jobs; rather, it can help them build ways of thinking and understanding that transcend specific tasks. This tension can be and usually is healthy. Academics in the employ of the military have to remember who they are while reminding themselves that their main readers and students are not going to become academics themselves.[7] Service members have to keep their missions in mind while forcing themselves to push beyond the bounds of those lessons that are obviously, and superficially, beneficial.

The tension becomes unhealthy when it drifts toward one pole or the other. In my experience, the military does a very good job of reminding academics where they work, and therefore there are few or no places in the military that have become purely academic. Lots of academics, on the other hand, struggle to make their case for the military. Right now, the pendulum has swung too far in one direction. And

6 See Kidwell 2005.

7 A good recent reminder is in Spiller 2006.

so the tension threatens to become artificial: because there are folks with PhDs around, too many in the military think service members are navel gazing, but they are only doing it Google deep.

What Can Be Done?

And so we come back to the issue of expertise. Expertise, more than anything else, is the stock and trade of academics. As a loose rule, most undergraduate degrees really only introduce the student to a subject area. My bachelor's degree is in History/Government, with a bit of a tighter focus on American history and government—enormously broad fields of study. Strong master's programs are more focused, and—through course material, the writing of theses, and connections with other graduate students—begin to expose students to the expanses of knowledge on all manner of subjects. My MA is in American history, with an emphasis on twentieth-century American political and military history, and my thesis focuses on Army rifle company commanders in the European theater in World War II. The work for that degree, on paper, might look impressive, but it was still a far cry from expertise in an academic sense.

Indeed it is a common misunderstanding that the next degree, the doctorate, endows its holder with thorough and expansive knowledge. The dirty little secret among holders of terminal degrees is the exact opposite. If the master's exposes students to extant material to be learned, the PhD lets them know exactly what they do not know. The logic is simple: the more focused an individual becomes on a narrow subject, the more that individual realizes how much there is to know about that one subject. I spent multiple years reading, researching, thinking, and writing about twentieth-century American military history for coursework, and a few more years reading, researching, thinking, and writing just about ethnic groups in the Army in the World War II era for a dissertation. It is not all that hard to imagine spending similar amounts of time and effort on countless other subjects. In short, we know what we do not know. By those standards, when I was done with my PhD, I was something like an expert on ethnic groups in the Army in the World War II era, and I had practiced some significant critical thinking on that subject. In the meantime, I did not become an expert on any other subject. Over the course of a career, PhDs can add other areas of expertise, but regardless, expertise, in this academic sense, is not really relative. In a relative sense—compared to the average—PhDs are remarkably well-read on a variety of topics (often related to teaching). But no one knows better than a PhD the difference between well-read and expertise.

Compare that experience to the military, especially the contemporary military. All service members switch jobs on a steady basis, so roughly every two years they go to new units with new tasks and responsibilities. This trend is especially pervasive among officers, who often make the point that just as they become comfortable and proficient in a position, they get transferred or promoted to another job. Officers, in particular, face another issue that runs directly counter to the academic experience: as they are promoted through the ranks, they go from the specific to the universal. For example, when Army officers are small unit leaders or commanders, they work in a narrow field of expertise based on a career branch: armor, infantry, artillery, aviation, and so on. As they move up to higher level command and staff positions, they become responsible for units and soldiers that cross specific branches. By the time they become generals, they take off their branch designators altogether. Along the way, their knowledge by necessity becomes more diffuse—they have to know about armor, infantry, artillery, aviation, and so on, but not to a level of expertise in each.

The contemporary military tries to mitigate this problem in a number of ways. There is a strong focus on education, both within and without the force. For example, Army officers who move up in the ranks get schooling and degrees from the Command and General Staff College and the Army War College (or equivalents), and a number of intermediate, non-degree granting programs along the way. They also are strongly encouraged to get additional master's degrees, usually from correspondence or distance education programs. In the past two decades, there has been a proliferation of such master's degrees among mid-grade and above Army officers. Obviously, those degrees are of varying quality and utility, but it is safe to say that even the very best can rarely do more than expose the students to knowledge without exposing them to true academic expertise.

Higher level Army commands also have a less formal method for dealing with gaps in knowledge: subject matter experts (SME). For example, staffs dealing with large problems in unfamiliar areas of the world will tab individuals to be designated SMEs on the variety of tasks and issues related to the problems and areas. Sometimes, they are able to bring in outside academics, journalists, or military experts to brief on the specific issue. Occasionally, an SME is an officer who has worked on one area for a long time in his or her career. But often the SMEs are staff officers who only have time and resources to do internet research on the subject. In any event, time constraints mean that the presentations and research are extremely limited. Despite the use of the phrase "SME," whatever such an expert can provide is almost always a far cry from what academics understand as expertise.

Frankly, this expertise gap is not going to go away, because it is impossible to change the physics of the situation. There just is not time and space enough for military officers to become experts, in an academic sense, on any of the myriad subjects with which they contend on an everyday basis. Moreover, academic-style expertise itself might be counterproductive. The nature of war is such that as they move up the ranks, service members shift emphasis from tactics to operations to strategy to advising military and national policy, without ever losing the need to keep an understanding of the other levels. By the American military's definitions, tactics are concerned with the planning and execution of battles and engagements, which are more limited in time and space. The operational level of war deals with campaigns and major operations that are planned, coordinated, and executed to achieve larger objectives in a theater or large area. Military strategy involves the coordination and application of military (and often non-military) resources to achieve theater or national military objectives. Advising military and national policy require officers to understand military capabilities and potential and place them in the larger context of American domestic, foreign, and economic policies.

Tactical efforts can and often do have higher level effects, and tactical leaders should know where they fit in a larger campaign, but for practical purposes, those working at that level can focus on that level. But working at the operational level requires a solid understanding of both tactics and strategy. Those at the strategy level must marry tactical and operational capabilities to policy objectives. And policy advisors have to think about where the full range of military affairs fit into the wider policy world. The required body of knowledge grows steadily, but time does not grow with it. What would a senior officer have to sacrifice in breadth if he or she spent the time to become an academic-style expert on any narrow topic? If one of the goals at higher levels is to coordinate across the levels of war, what vital aspects for coordinating might be lost if an officer spends too much time on one subject?

The all-too-common solution is to get academics to teach and write about everything and anything that might be relevant. The addition of new material without the removal of something else means that everything becomes more shallow, even if academics are doing the teaching or writing. That solution is how we get to navel gazing Google deep—the mere presence of academics does not mean that their expertise will transfer. Another solution, particularly in the realm of education, is to teach how to think. As far back as 1961, eminent historian Michael Howard encapsulated "how to think" under the far more elegant entreaty that "the true use of history, military or civil, is … not to make men clever for next time; it is to make them wise for ever" (1962: 8). If the military audience needs to know so much that it is impossible to absorb it all from any combination of specific readings or educational programs, then perhaps the best that can be hoped for is that they learn the ability to think critically. That way, when they confront all of the situations for which their reading and education did not directly prepare them, they will have the capacity to work their way through the problem. Since most advanced academic study is about critical thinking—whether the academics understand the mechanics of critical thinking or not—then it follows that engaging in such study will help develop the ability to think critically. This approach, while appealing to academics, has its own problems. Being able to think is essential, but not just in a general sense. It cannot be, because the pull toward academic arcana, navel gazing, is too strong.

Despite these obstacles, the effects of the expertise gap can be mitigated. The military can make better use of academics (and academics can better serve the military) without swinging all the way to an academic model. Service members do need to know how to think, but as practitioners their thinking must have a domain of applicability. Even at the highest levels of the military profession, even in a complex world where policy, politics, diplomacy, economics, culture, and communication all intersect with their everyday lives, the military is still the military. What they require is to be able to think about military affairs, a broad but not universal domain of applicability. Michael Howard began to make this point in a very Clausewitzian way in regard to military history: "despite the growth of 'war studies' and 'war and society' … at the center of the history of war there must lie the study of military history—i.e., the study of the central activity of the armed forces, i.e., *fighting*" (2006: 20).

The task for the American military, as a profession, is to draw some stronger boundaries around their domains, especially as those domains relate to academic issues. This is no easy requirement, because of the long and tendentious history of American military professionalism. From the very beginning, the United States has had a mistrust for the type of militarism that most clearly defines the military realm, the kind that accompanies standing armies and professional general staffs, because that sort of Cromwellian or Prussian militarism threatens free government. Within that context, American military professionalism necessarily developed with an eye toward assuaging worries that the military would become too coherent, exclusive, independent, and therefore threatening an institution to American ideals. Ironically, even as some have argued that the military in recent years has become too professional and isolated from the civilian realm, the force has also expanded its domain of applicability. This trend probably has something to do with the steady expansion of missions and responsibilities for the military as the United States has gone from continental power to regional hegemon to global superpower. In all of these eras, the intellectual aspects of professionalization have been challenged by a military culture that tends to (falsely) separate thinkers and doers, and overwhelmingly favors the latter over the former. Regardless of the specific factors involved, the effect is that the American military has never quite settled on the standards required of a member in good standing in their profession.

But just because the reality of the specific characteristics of the United States and its interactions with the world make it harder to define professional standards, that should not stop the effort. There are many ways to limit the field, especially in regard to academic issues. When it comes to academics, most of the relevant standards are intellectual, and therefore relate more to education than training, organization, or actual combat. Academics interacting with the military are most properly placed in educational relationships—teaching at schools or writing studies to foster learning.[8] More specifically, even in today's complex environment, as a military professional moves through the ranks he or she still has clear constraints on responsibilities. A senior officer working at the strategy level has to understand policy, politics, diplomacy, economics, culture, and communication, but primarily insomuch as those subjects intersect with military strategy. For the purposes of an academic teaching or writing for such an officer, the domain of applicability is

8 The advising role of academics, including working in the field on what are now called Human Terrain System teams, is dealt with by Carol Burke in this volume.

military strategy. If the officer in question comes from the Army, for example, then the domain can be further refined to emphasize the role of land power in military strategy. The same thought process could be applied wherever academics work with the military.

Within such domains, it becomes easier to set baseline intellectual professional standards. For example, every Army major should have read certain books and engaged certain concepts, and then been tested in some way to show that the minimum has been met.[9] The domains and standards would give focus to academics working with the military, allowing them to apply their skills in a more useful manner. This recommendation is not about lists of learning objectives for the academics to use in teaching or writing; it is about using their most valuable asset, their understanding of expertise, to push military service members to think critically at the appropriate domain of applicability.

The tension between academic and military culture for academics working in the military will never go away entirely. Academic expertise will never be an appropriate standard to expect for members of the military. But narrowing the expertise gap is something we must constantly strive to do, because navel gazing Google deep simply is not good enough.

9 All of the services have reading lists, but there are no functions to force anyone to engage with them. A slightly more sanguine view for the current state of applied history can be found in Lynn 2008.

Chapter 10
Professors in the Colonels' World

Daniel J. Hughes

This chapter addresses some of the issues facing civilian professors teaching in a senior military school, in this case the Air War College at Maxwell Air Force Base, Montgomery, Alabama. The virtual pendulum that swings back and forth between the rewards and the challenges of teaching in such a school reflects a deep divide between academic and military cultures. Just as pendulums provide no bridges between opposites, many civilian faculty members find no easy balance between their approach and that of their military supervisors. The following text reflects the author's subjective conclusions based on 30 years of employment with the Department of Defense, the last eighteen of which were at the Air War College. Because every broad characteristic of such institutions has its exceptions, one must be careful to recognize the limitations of the generalizations offered below.

The School

The Air War College is the senior service school of the Air Force and part of the "Air University." It is a counterpart to the war colleges of the Army, Navy, and Marine Corps. The Air University commander, an approximate counterpart to a university president, is a lieutenant general (three stars), while the commandant of the Air War College is a major general (two stars). The deans at the Air War College have always been either officers or retired officers, without exception, some with doctoral degrees and some without.

The curriculum of the Air War College consists of mandatory core courses and a series of electives along with a capstone war game and a variety of so-called distinguished visitor lectures. These core courses primarily address strategy, national security decision making, the international security environment, leadership, and war fighting. Officers teach in all these courses, while civilians, with a few exceptions, teach in the first three. Both civilians and military faculty members teach electives, which are supposed to allow the faculty to teach specialized courses to give depth to some core areas. The Air War College offers an MA degree in "strategic studies" accredited by the Southern Association of Colleges and Schools.

The Faculty

Like the other senior schools of professional military education, the Air War College has a mixture of civilian (25-30 percent) and military faculty members. Air Force officers constitute about 60 percent of the military faculty, which also has officers from the Army, Navy, and Marine Corps. This chapter refers to the military faculty as "the colonels" and uses the masculine pronoun because very few women serve on the faculty. These senior officers are remarkably similar, sharing many strong and weak points across service lines.

The colonels as a whole have little in common with their civilian colleagues. The colonels may be effective officers, but that is not why they are on the faculty. Doers rather than thinkers, they are practical people who have succeeded in numerous military assignments. The services do not assign them to the Air War College because they are particularly suited to academic work. Indeed, most do not perform much traditional academic work. Many are of average intellect and have substantial weaknesses in speaking and writing. Most are quietly, and some even openly, contemptuous of intellectuals and regard them as guilty of the egregious offense of being "liberal," a four-letter word in the US military.

A very few of the colonels have doctoral degrees, usually from on-line or other schools at the bottom of the academic world. A few hold doctoral degrees from reputable graduate schools. All have MA degrees, some respectable but many also of a shadowy nature. Most share none of the ethos of the professor or of the civilian university and have little understanding of what scholars do. They rarely publish. They rarely lecture, even at the War College. Indeed, at the end of a 30-year career, the typical military faculty member, regardless of service, proclaims his inability to lecture, even on strictly military subjects.

It should be noted that an assignment to the Air War College is not an asset to the colonels' careers, which in most cases are almost over when they arrive. The Air Force considers these officers unsuitable for further promotion and has not elevated a single one of them to brigadier general in at least the last eighteen years, and probably longer. Many colonels try to avoid service at the Air War College and at the Air University generally. Most regard an ROTC assignment as a better way to end their careers, an attitude enshrined in the fact that there is a competitive selection board for assignment to ROTC duty but not to the Air War College.

The Air Force routinely sends colonels who are out of favor to ride out their final years at the Air War College. One had been relieved over the 2007 nuclear warhead incident at Minot Air Force Base. Another had been held responsible for the crash of a very expensive airplane. One was emotionally disturbed, having claimed to have suffered all kinds of abuse while on embassy duty. Others remain on the faculty for many years because no other Air Force agency wants them. These are the "Maxwell homesteaders."

One also encounters that growing group of people known as "hybrid" faculty or "colonel-doctors." These have obtained doctoral degrees, at full pay and all expenses, then serve first as colonels and then as civilians upon retirement. A wide variety of graduate schools give them favored treatment so they can complete their degrees in as little as three years. The Army War College leads the way in this program of selective in-breeding, but the Air University is attempting to catch up in all its schools. These ersatz civilians normally are quite deficient in the breadth and depth of their knowledge but lack nothing in self-confidence. They usually acquire none of the ethos of academe. They should not be regarded as civilian professors in the sense that Congress has intended when establishing guidelines for percentages of non-military faculty in the military schools. Management is impervious to the extent to which these colonels undermine the cultural and intellectual diversity of the faculties in these schools, but welcomes them because they are easily controlled and are fully socialized into the military approach to education and training. For the Air Force, these "colonel-doctors" constitute the ideal "civilian" faculty members. While career officers, they have sufficient "academic stink" to legitimize their civilian faculty status. They are "team players."

The case of one of these colonel-doctors is particularly revealing. This person, retiring from a similar position at one of the other war colleges, applied for a job as a department chair at the Air War College. During the public portion of the interview, he reached into his briefcase with great flourish and triumphantly pulled out three books that he announced would be the core of what he would use to teach strategy. When asked to compare the books' views on a particular subject, he had to admit that he had not actually read any of them. This same person offered the opinion that a professor's fund of knowledge peaks upon finishing graduate school and declines thereafter. As ridiculous as this episode was, the dean, then also a retired colonel-doctor, nevertheless wanted to hire this person as a department chair.

The civilians, with some exceptions, are very different kinds of people. Almost all have earned PhD degrees from the middle tier of graduate programs, and some have credentials from top schools in their disciplinary specialties. Only a few have had any military experience, and this number is declining as the baby boomers take their Vietnam experience into retirement. Only a tiny minority has left academic positions to join the faculty, and most have never had a regular faculty appointment at a college or university. For most, a position at Air University is their first real job after graduate school or their first permanent position after a string of temporary appointments. The civilians seek to continue their academic pursuits while working in the bureaucracy and steadfastly hold onto as much of the academic ethos as is possible under the circumstances. They try to maintain their specialties, to conduct research, and to write scholarly books and articles in the manner expected of academics. This article refers to them as the "professors."

Official differences in work priorities officially endorse the divisions within the faculty. Like most academic institutions, the Air War College expects its faculty to be active in teaching, publication, and service. The top priority for both the colonels and the professors is teaching. For professors, research is second and service is third, while these are reversed for the colonels. In fact, the College does not expect the colonels to publish anything, and most live up to that expectation. On the other hand, most professors would not consider themselves good candidates for commanding military units, and recognize that colonels should not be expected to conduct academic research and writing to any considerable degree. For the most part, academically inclined officers in blue uniforms teach at the Air Force Academy, which is more important to the Air Force than is the Air War College.

The professorial office, that most personalized of academic places, illustrates the cultural gap between the professors and the colonels. Upon entering an office, the knowing observer instantly recognizes if this is the abode of a professor or a colonel. In a professor's office, one finds walls of bookshelves, stacks of papers, research notes, and drafts of manuscripts, loose books, and so forth. The colonel's office has only those books required for the courses he (or in very rare cases she) teaches and little if any of the normal trappings of the academic existence. Instead, the typical colonel's office is adorned with the accoutrements known collectively as the "I love me wall." All around the office one finds evidence of the colonel's past success and comradeship. These include pictures, certificates of awards, plaques of every kind, statues or similar icons of the warrior class, flags, old scarves (if one is a fighter pilot), parts of airplanes, approving statements of grateful subordinates and the ubiquitous collection of "coins." These, not academic materials, reflect the colonels' past, present, and future.

The Students

Drawn from all branches of military service, the students are junior versions of the colonels. A few, in fact, are colonels. Beyond rank and age, there is little meaningful difference between the colonels and the military students. The student body's 45 foreign officers provide the only real diversity (gender excepted). The Air War College also has about fifteen or so civilian students each year. Most are employees of the Air Force.

The Air War College may be the only full-time graduate school in the country in which a large part of the student body does not wish to be present. Students attend because graduation almost guarantees (98 percent or more) promotion to colonel. This author's informal and unscientific questioning of students suggests that at least 50 percent of them have no interest in attendance and would not attend if they could otherwise be certain of advancement to colonel. I have never had a group of students deny that most would decline to attend given the above guarantee.

This produces a very unusual situation in the classroom. Many students attend classes without enthusiasm or any intellectual curiosity beyond concrete course requirements. Many refuse to read the assigned material, especially late in the academic year. Many complain they cannot complete the assigned pages (80 maximum) for each class. In all too many cases their attitudes are similar to those of disinterested undergraduates. Complicating matters further is the school's leadership's practice of welcoming each new group of students with the admonition that this year is a time to "recharge their batteries" and to "get to know their families."

Differing Expectations

One of the most fundamental sources of friction has to do with expectations. The professors want the core courses to exhibit some of the rigor they encountered in graduate school. Students and the colonels generally stress practical applications because they have a bottom-line approach that demands answers rather than unresolved difficult questions. Reductionism becomes a staple of the student approach, while the professors attempt to teach the complexity of international relations, of strategic issues, and of other related subjects. Students and many of the colonels want to stress subjects that contribute directly to the jobs students are likely to have upon graduation, while the professors are more interested in broader questions and less definite answers. Students cling to those staples of the military apparatus, the ubiquitous templates and checklists, while their professors seek to raise the student comfort level with complexity and ambiguity.

Students, especially Air Force students, have little interest in theory and history, which they frequently regard as having been made irrelevant by advances in technology and military capabilities. Professors stress continuity in basics and the use of theory and the past as guides to the continuity in politics and human behavior. Professors frequently regard the students as willfully ignorant of contexts, while students and colonels frequently regard the professors as dismissive of or improperly inattentive to ongoing developments in military technology and capabilities. To use an old cliché, students and colonels have a tendency to stress *what* they need to know in an immediately practical sense, while the professors are more interested in *how* to think about issues of policy, strategy, and so forth.

Colonels and professors have little in common, but the reverse is the case for the colonels and the students. In addition to their uniforms, their overwhelming Republican Party affiliation, and their addiction to Fox News and Rush Limbaugh, the colonels and the students share a common anti-intellectualism that has no parallel in academic institutions. Most have little if any respect for learning that does not arise from practical military experience. Most apparently rarely read a newspaper, except perhaps for the *Wall Street Journal*, and seldom read books beyond those snippets required for courses. They have an infinite familiarity with *American Idol* and malicious Internet gossip about Democratic Party politicians, together with a devotion to Right-wing political slogans or fundamentalist religious propositions that is impervious to facts and reason. They are genuine devotees of the Internet. Their collective image of the academic community reduces that complex group to a simplistic caricature evolving around the hated "tree huggers, do-gooders, and socialists." Ironically, they who live in the most *de facto* socialist and bureaucratic large community in the United States regard themselves as the vanguard of rugged individualism, independence, and capitalist entrepreneurship.

By and large, there are no real academic standards, a fact to which new professors quietly object but which they, like the older hands, eventually accept with resignation tempered with dark humor. The effort of a previous dean (a retired officer) to address this problem by leading the students in chants of "Rigor! Vigor!"

had little short-term or long-term impact. In 2009, the war fighting department gave the grade of "A" or "A-" to more than 97 percent of the students in its core course. All the instructors in that department are officers or retired officers, some of whom also give extreme proportions of high grades in electives. A few civilians hold the line individually but some do not. For academic year 2010, the dean reinstated an old "goal" of no more than 50 percent "A" and "A-" grades. A professor who assigns low grades or too much work in an elective endangers future enrollments, thus jeopardizing continued employment. As a result, even in the electives academic standards are frequently, but not always, quite low.

For decades, the Air War College has appended one element to its academic program that is largely unknown to academic graduate institutions: remediation of course failures. Students who failed examinations (which rarely happens) or who produced unacceptable papers traditionally had the absolute right to re-accomplish the task because the officers are viewed as expensive assets that cannot be allowed to fail. Virtually 100 percent of those who fail the first time pass the second effort. The insignificance of performance in school also plays a part, since the main point is to graduate, not to excel. Assignment after graduation bears no relationship to school performance. Those whose writing was very poor once found themselves in remedial English programs, but these have fallen out of favor in recent years, as PowerPoint has overtaken the requirement for English literacy in the Air Force. For 2010, the commandant has tried to eliminate remediation, but it remains to be seen if that will merely reduce the already tiny number of low grades.

At this point the institutional climate intervenes to erode the academic environment even further. Professors are reluctant to give grades of "C" or less because the remediation process punishes them rather than the failing students. In fact, professors frequently consult with each other to determine if anything is to be gained by forcing a student to repeat an assignment. If the student lacks the ability to improve his performance, there is no reason to force him or her to do it over since the end result is the same as giving a meaningless "B" or "B-" the first time.

There is some justification for regarding the students as too important to fail. The military has invested substantially in their careers and in its senior schools. The services, moreover, need officers with at least some exposure to the topics taught at the senior service schools. A student who fails to graduate probably would not be promoted, would not be qualified for joint assignments, and would become a retiree in a relatively short period of time. Since the services invest at least $300,000 per student per year, to say nothing of the incalculable investment made in every officer prior to senior school attendance, the reluctance to cause or even recognize failure is understandable in some quarters. The tension between academic standards and the service's investment remains a problem that is hopeless but not serious. The taxpayers, after all, are very generous people.

Sources of Institutional Friction

Although the accreditation agency in theory demands a central role for the faculty in institutional governance, there is no such thing at Air University. There is no faculty representative body and the faculty normally has no role in decisions on policies that fundamentally shape the curriculum. As in any military institution, the commander has absolute power to make decisions from which there is no appeal. As a result, the work environment for professors parallels a labor-management scheme rather than a model of faculty governance. In this particular case, unlike most labor-management models, there is no representation or negotiating opportunity. Rather, everything depends on remote application of civil service rules, which are very weak under the special provisions of Title 10 employment in the excepted service.

Ultimately, this is a rule of men, not of laws. The College's rules and procedural guidelines, called "operating instructions" (OI; now with new names, but basically the same items), govern most major activities in extraordinary detail and seem to be a mechanism for creating a very orderly organization. In practice, though, an OI has value and provides guidelines only when management wishes to use it to enforce its desires. When that is not the case, management changes the suddenly obsolete OI, frequently without even informing the faculty and sometimes retroactively. OIs thus govern labor but not management. This is, once again, entirely in accordance with the needs of a military organization and produces only a harmless cynicism among some professors. The colonels seem quite accustomed to the arbitrary exercise of authority from on high, and the professors soon come to accept it or leave. It is a sign of their loyalty to the institution that they do not routinely warn job candidates of this issue during the hiring process.

One of the most serious challenges faced by many professors is the loss of their treasured academic specialization, which, after all, is at the core of one's professional identity. Core courses are broad surveys, largely at the undergraduate level, covering the same subjects and readings across the sixteen or so seminar groups on the same day. These simplified courses introduce students to a large number of topics but treat none in any depth. For example, a course in "international security" consisting of 15 lessons might have the following topics: a class or two on US interests and one class each on the major geographic areas or issues (Europe, Central Asia, South America, Middle East, China, globalism, and so on). A specialist in each of these areas selects the readings and designs a lesson, which the other 15 instructors also teach. Each instructor thus has to learn enough to teach so-called graduate-level classes on all these areas.

Most professors thus become generalists, repeatedly teaching courses that are assemblages of basic approaches based on undergraduate-level readings. If the department lacks a specialist in a topic in need of an instructional period, someone acquires instant expertise. Since the resulting superficial treatment rarely goes beyond a handful of introductory readings subject to the standard page limit, core courses rarely rise to the level of real graduate education. Superficiality, in short, is the coin of the realm in such courses.

An accompanying extreme degree of standardization keeps most core classes at a uniformly low level in approach and content. In core courses, all instructors use the same "scope sheets" as the basis of the classes. A scope sheet sets forth the lesson objective, the readings, and the official questions to be used by faculty and student preparation, and the ubiquitous military learning system of "desired learning outcomes." The faculty member who develops the "scope sheet" also prepares a "faculty guide" and conducts a workshop to prepare the rest of the faculty for the class session. This is to ensure conformity to what the College and outside DOD agencies have determined the students need and to support faculty who do not have the ability to design or teach such classes. This is quite in keeping with the Air Force's "training" approach to education, but is foreign to professors, whose model of academic activity stresses individual responsibility for the substance and form of what they "profess."

The military schools simply cannot trust their instructors to develop core courses on an individual level. Regardless of how foreign this might be to individual professors, there are good reasons for it, principally the significant percentage of the faculty which is unprepared to teach such courses. The colonels embrace standardization, since most simply cannot design courses on their own. Most have great difficulty in creating individual class periods, let alone entire core courses. A real revolution in military education and official outlooks toward the expected outcomes would be necessary to change this standardized approach. Academics should avoid such employment if they regard this as unacceptable.

The difference in outlook between the colonels and the professors is obvious in another aspect of institutional control common to bureaucracy but entirely foreign to the academic workplace: presence in the office. Although enforcement waxes and wanes as personalities of military superiors (commandants, deans, department chairs) change, the enduring rule is that the professor is supposed to be in his office or some other official place during the "duty day." This is standard in most places of the federal bureaucracy and for good reason. Most bureaucratic work requires routine access to the instruments of government. In the military academic environment it is neither necessary nor desirable, but adherence to it does prevent non-academic civil service employees from complaining of special treatment for instructors, who are in a different category in the civil service. But there is another reason for the practice.

Any concept that professors perform work in their homes, let alone that a home office might be more productive than the government-issue facility, is entirely foreign to most officers. Officers apparently do little or no work at home, which is where "the wife" and the children are and to which he (she) returns when work is done. Their view is that if one is not at work, one is not working. Government offices, for their part, are very unproductive places in many cases. Computer security constantly restricts access to many standard web sites. Air University is particularly incompetent in managing access to the Internet and to email. Its "fire walls" frequently deny email access to all sorts of communities, including other military and even other Air Force agencies. The addition of a second computer system using the EDU domain has provided some relief, but many problems remain. "Exercise" notifications frequently disrupt productive work. Technology has enabled bureaucrats and supervisors to disrupt any work day with all sorts of emails, to which they sometimes expect short-term replies. The constant traffic in and out of work areas, hallway conversations, and so forth provide a steady stream of interruptions.

Even within the official "work site," school authorities find further ways to illustrate their complete lack of understanding of the nature of academic work. At one point, the Air War College tried to enforce a standard office configuration that would have given everyone the same set of furniture, including a single bookcase. The faculty overcame this nonsense, the product of a particularly foolish and ignorant commandant, but not without difficulty. During a period of renovation, the "leadership" attempted to force the faculty to get rid of all those worthless books, notes, and other "clutter" that was not needed for being on the faculty at the Air War College. At least one faculty member was forced to put material into a common storage room in another building, from which some of it disappeared. By convention, doors must be open, thus creating endless opportunities for interruptions. On more than one occasion colonels have contemptuously asked the question, "how many of those books have you read?" or something similar. Others pick up research notes or other papers with the derogatory comment: "what is this?" These comments do not always reflect personal hostility on the part of the colonels. Rather, they are indications of the gap in basic perceptions of the nature of work at an academic institution.

The terms of employment at the various service schools provide further indications of the non-academic nature of these institutions. This particularly is the case at the Air University. Beginning around 1990, the Air University instituted a tenure system similar to those at most universities, but then abolished it after 2000. Instead, it offers initial appointments of three years and subsequent renewals of periods from one to five years. Those persons who had tenure were allowed to keep it, but those who had been hired into tenure-track positions were given a gruesome choice. Air University schools offered renewals longer than a single year only to professors who agreed to give up their tenure-track appointment. Since a single-year renewal suggested termination at the end of that period, every untenured professor eventually chose the longer appointment.

The result is that, except for about a half-dozen individuals, the professors at Air University are in the category of the proverbial "permanent assistant professor." Every two, three, or five years, if they are fortunate, they have to re-apply for their own jobs. There are no firm standards for reappointment, and beyond the initial re-appointment the faculty has no voice in renewal decisions. In 2009 Air University announced it would renew appointments no sooner than 120 days before the old term expired. This means abandonment of the traditional academic policy of one-year's notice of faculty dismissal. One professor suggested that the Air War College should change the title of its professors to "teaching employees," the acronym for whom would be TEMPS.

The normal friction between faculty and administrators takes on a different character at the Air War College because of the nature of the supervisors. In the academic world, administrative authorities normally have very solid backgrounds of achievement as professors before they become chairs, deans, and so forth. At the Air War College this is not the case. Most superiors have little or no background in academic work. Rather, they are successful "operators." This finds some justification in the nature of the school and its responsibilities within the Department of Defense. Nevertheless, it is a source of continuing friction because members of the "senior staff" normally have no academic experience and act in a manner consistent with military units but not with a graduate school.

The amusing and irritating behavior of ignorant bosses frequently illustrates the gap between professors and colonels in administrative positions. One vice commandant frequently spoke of "Saudi speakers" because he did not know that Arabs speak Arabic. A chair of the Department of International Security made frequent reference to "Muslim speakers" for the same reason. Another senior military administrator used his authority to force adoption of an elective on Civil War soldiers and wore his re-enactment uniform around the building. Another consistently confused "monogram" with "monograph." Most have no clue of the differing natures of publications and do not know the difference between a vanity press and a respected university press. And so it goes.

In this author's eighteen years at the Air War College, only three civilians have served as departmental chairs, and those only for short periods of time. Chairs are normally colonels who have never served on any faculty before coming to the Air War College and have at most a year or so as faculty members before becoming heads of their departments. Most have never conducted any serious research, have never written an article for a journal of any kind, and possess no educational background to give them familiarity with the literature of the disciplines of their professors. Many have never put together courses or even developed individual class sections and have no idea what is involved in doing so. Most have very authoritarian views of their positions and do not regard themselves as colleagues of their subordinates. Many have no understanding of the differences between an educational institution and a military unit. They are truly amateurs, sharing neither the academic backgrounds nor the academic ethos of their civilian subordinates.

Civilians frequently have a hard time adjusting to another aspect of teaching in these military schools: the burdensome array of administrative requirements attendant upon any governmental bureaucracy. Many of these are necessary. The Air Force requires its civilians to take recurring training in force protection, information assurance, records management, privacy act, suicide prevention, combating trafficking in persons, sexual assault prevention and response, operations security, whistle-blowing laws and regulations, all kinds of safety concerns, and so forth. Exercises frequently disrupt work days, as does compulsory attendance at all manners of lectures and other events made mandatory at the whim of the commandants or deans. The unavoidable fact is that teaching in a military school carries with it an array of distractions not encountered in most civilian universities, including numerous "chamber of commerce" speeches by senior generals. The only AWC professor ever to possess fluency in a Middle Eastern or Central Asian language was fired at least in part because he refused to attend some of these sessions, which he regarded as a waste of time. Needless to say, most professors find time for them.

Anti-Intellectualism

Professors at the Air War College labor in an overtly anti-intellectual environment that reinforces the basic tendencies of the students to disparage all but practical experience. The generals in charge of the Air War College and Air University frequently directly contribute to this poisoned atmosphere. One Air University Commandant, for example, routinely used nearly every address to the students to brag about what a poor student he had been while an undergraduate at the Air Force Academy. He never had any summer vacations, he explained, because he was always repeating courses which he had failed during the year. He was proud to have finished near the bottom of his class at the Academy. It never occurred to him how incongruous it appeared for the president of a "university" to have been a poor student and to barely have graduated. Nor it seems, did he ponder the curious fact that at the military academies non-athletes could remain on very generous scholarships (all expenses plus pay) while failing to maintain a 2.0 average. Rather, he used his speeches to remind the students that he had attained three-star rank because of his "leadership" abilities. He went on to attain four-star rank, became the boss of the Air University commander and was in charge of nearly all education and training in the Air Force. His message, that academic accomplishment is insignificant, was not lost on many students. Nor should it have been, since he was correct.

This Air University Commander was by no means exceptional in his open disdain for learning. The general officer corps of the Air Force is aggressively anti-intellectual in many of its presentations at the Air War College. One four-star general bragged (in February 2008) that he had been a student who did not like studying and "learned by asking rather than by reading." Numerous lower-ranking generals who formerly were students recall how they slept through lectures and remind the students that "it is only a lot of reading if you do it." Indeed, many refer to Jones auditorium, the large lecture hall, as the "big blue bedroom." It should he noted that few three- or four-star Air Force generals make such statements because the service consistently sends those destined for high rank to one of the other war colleges, in particular the National War College in Washington D.C.

Anti-academic posturing and anti-intellectualism on the stage has long been an effective way of pandering to the students by disassociating the speaker from the discredited academic world of the civilian professors. A retired admiral who joined the faculty as a civilian joined the attack on academic pursuits. For years he scoffed at "political science mumbo-jumbo" in nearly all his lectures in the big blue bedroom. He did this in front of the students, colonels, and the professors, about half of whom have degrees in political science. The foundation of his expertise, of course, was his practical experience in the Navy. Beyond that, he lacked any academic credential that would normally be necessary to qualify someone for a graduate faculty appointment. He was otherwise a congenial colleague, but was so steeped in the anti-intellectualism of the US military that he was oblivious to the way in which he belittled much of the civilian faculty in front of the students and the military faculty.

One Air War College commander, regarded as one of the rising intellectuals among the Air Force's generals, frequently exhibited his complete lack of understanding of the academic world even though he held a "doctorate" (not a PhD) in engineering from a private university in a Midwestern city. When it was that he actually attended any classes there or how exactly he received his "doctorate" was not clear, since his official biography listed no period of time during which this might have taken place in the course of his Air Force career. On numerous occasions he repeated his firm belief that writing academic papers was not difficult because finding his dissertation topic had been much more difficult than actually writing the dissertation. Research and writing, he always continued, was "easy." Needless to say, his own academic writing is very difficult to find. He could universalize his relatively unique experience without fear of contradiction not because he commanded the respect of the faculty but because his authority could not be challenged in public. In this case as in so many others, the hierarchical nature of the military organization forced the professors quietly to accept this belittling of their work in front of the students even as they were expected to continue it as a condition of their employment.

The disdain for academic pursuits emerges from a perennial cooperative enterprise of all the war colleges: Jim Thorpe days. These consist of three days of athletic competition among the student bodies of the senior PME schools, held at the Army War College in April. The faculties provide coaches, making these schools among the very few graduate schools (if there are any others) whose faculties serve in this capacity. The competition, which serves no readily apparent useful purpose for future senior officers, closes the schools for three days. One Air War College commandant, by general acclamation of the senior faculty the most intelligent of the entire group of commandants since 1996, cancelled it during his single year at Maxwell. Usually about 100 AWC students attend. This is, of course a questionable use of government funds, but the amount is not terribly large, since the students ride busses for the long trip to Carlisle (15 hours or so). For this chapter's topic, however, the real issue is what goes on for the 150 or so students who remain at Maxwell.

The Air War College punishes these students for not participating in the games. For years the faculty used the term "punish" as a bit of humor, but in 2009 officials at the War College used this term to explain the program to the students. The punishment consists of additional courses, called "enrichment electives," which are mandatory for the students. Professors volunteer or are dragooned into teaching courses that lack sufficient value to be taught in the regular elective program, but which suffice to avoid giving days off to those students who do not wish to pretend to be athletes or cheerleaders. Designing pseudo-academic courses to punish students for not participating in sports competitions is probably unique in the annals of American graduate education and is a clear statement of what really counts at the Air War College.

Academic Freedom

The tension between academic culture and the military bureaucracy finds yet another manifestation in the area of academic freedom. The Air University subscribes to the principle of academic freedom along the lines established by the American Association of University Professors. Proper formal statements are detailed in the appropriate governing documents and the Air War College probably has more academic freedom than many would expect. Professors and students are free to express their opinions on topics under discussion in the classrooms, although military persons must abide by the restrictions of the Uniform Code of Military Justice. This criminalizes contemptuous statements about superiors, various governmental officials, and members of Congress. All written products of the faculty and students must undergo review by the Public Affairs office and a security review. The former is designed to prevent misrepresentations (intentional or innocent) of Air Force policy. It is not designed to prevent criticism if the policy is correctly stated. The security review is to ensure that no classified material accidently finds its way into open sources.

A few years ago, a professor tenured at a civilian university turned down a job offer at the Air University because he could not have confidence that a faculty unprotected by academic tenure would have academic freedom. This comment correctly identifies the real issue in academic freedom at all of the Air University: self-censorship to avoid controversy or retaliation. Professors who must reapply for their jobs every few years can have no assurance of protection and rarely write anything really critical of the Air Force and its policies. Just as potentially controversial writings are rare, so are flagrant violations of academic freedom by the school's hierarchy. Nevertheless, such cases do exist. A commandant once summoned this author to warn against publishing a controversial article in the Air University's house organ, *Air Power Journal*. Another faculty member recently published a critical commentary on a well-known retired colonel's article on Clausewitz in the Air University's new journal, *Strategic Studies Quarterly*. A number of retired officers, including at least one general, wrote to the editor demanding action (including an apology) to redress what they regarded as an affront to the retired colonel. Fortunately, the Air War College's dean, a retired lieutenant colonel, defended the faculty member. The episode nevertheless demonstrates the kinds of pressure used to restrict writings critical of official policy, institutions, or officers. In another recent case, senior officials gently cautioned a professor not to submit a potentially offensive article for publication because it was critical of a senior Air Force general.

More in evidence is the so-called "Blue Dart" program, created by the same Air University commandant who bragged about his poor academic performance at the Air Force Academy. This policy requires all faculty members at Air University to write one "Blue Dart" per year. A Blue Dart is an "influence piece" of about 750 words (none to contain more than eight letters according to the original instructions) discussing "some aspect of air, space and cyberspace." It should "draw attention to a subject of importance to decision makers and influence leaders in this country and beyond." While in theory a Blue Dart may disagree with an Air Force position, the intent, as stated numerous times by public affairs officers, is to "get the Air Force message out." Each faculty member must, as a part of his official work plan, send such a Blue Dart to the Public Affairs office. That office judges its utility and submits the Blue Dart to any publication it so desires, regardless of the faculty member's preferences and even without his knowledge. This use of the faculty as a source of propaganda for the Air Force's outreach program met with nearly universal contempt on the part of the faculty, but of course also produced compliance (frequently grudging). Some avoided this threat to their academic integrity by intentionally writing pieces with a tone such that they would not be published. This counter-productive result seemed to escape the Air University leadership, which continues to press this requirement year after year.

Paternalism

The US military, entrusted with the lives and welfare of many citizens and foreigners, is very paternalistic, probably more so than any other institution in the country with the possible exception of a few cults. One commandant recently proposed sending letters to the parents of new civilian professors to congratulate them on joining the faculty. This harmless kind of foolishness is merely the tip of the big brother iceberg just beneath the surface. Much of this is merely an extension of the services' concern for the large number of young adults under their care. Other measures are part of the processes of socialization and control. Superiors routinely withhold information on all sorts of changes in administrative or academic programs with the later explanation that they did not want the faculty to worry. One department chair acquired the nickname of "dad" because of his tendency to treat the civilian faculty like one would children or very junior enlisted soldiers.

Ideology and Politics

Overall, senior professional military education exists in an atmosphere of very conservative views in almost every area of public and foreign policy. The colonels and retired officers in civilian positions easily constitute between two-thirds and three-quarters of the faculty in the senior schools and as a group are ultra-conservative. The students are just as conservative, perhaps more so, as the self-selecting process of the all-volunteer force has made the military an uncomfortable place for anyone who is on the left or even in the center of American politics. This author has known only two colonels who professed to be Democrats, and these always tried to keep that a secret and always spoke in hushed tones when discussing political matters. The professors, like much of academe, are moderately liberal and are much more likely to be to the left of Newt Gingrich on fiscal and social policy than are the colonels and the students.

This hostility extends to liberals in Congress, the media, and other institutions all the way to the top. Many colonels continue to demonstrate their famously open hostility to former President Clinton and naked contempt of President Obama. Shortly after the inauguration, Air Force officers widely distributed an email portraying the new President as a laughing pig ridiculing any idea of a military pay increase. An Army student at the Air War College sent an email (April 21, 2009) commenting on a very pro-military article in the *New York Times* about a successful platoon-level action in Afghanistan. The officer wrote: "That's ok, (sic) it's in parallel to the new administration's policies on telling the enemy everything we can so he won't get hurt." The blunders of the Bush administration are, to paraphrase Cicero, passed over in silence.

Even though most professors attempt to avoid discussions of politics and controversial social questions, the ideology of the military definitely impacts the nature of education. In one discussion (in a class on civil-military relations in 2008) the author made reference to an unidentified court case allowing absentee ballots to reverse the election-day results. One of the students (a colonel) immediately interrupted and angrily blurted out in a quite hostile tone: "Why don't you just admit that Bush won the election?" He could not wait until the professor could explain that these results were from a county election in Kansas in 1966, chosen because it would not be controversial. This underlying defensiveness about President Bush is representative of the Republican Party orientation of the students, who regarded most civilian intellectuals as liberal enemies and many of whom aggressively look for opportunities to make their views known.

Controversial issues do come up, primarily in discussions of national resource allocation, civil-military relations, and gays in the military. Many of the students are simply not receptive to alternative views on certain issues and react strongly when presented scholarship that is not in accord with their own opinions. They maintain that much of the national polling is tainted with liberal bias, but place complete faith in similar polls that show great public respect for the military. Quite a number are biblical literalists who deny evolution and attribute homosexuality to a conscious choice of a sinful life-style. Many will not consider global warming as anything but a liberal plot. Others dismiss it because it is associated with Al Gore, whom many hate with a consuming passion. Many use any discussion of social and economic problems to pillory anything to the left of Milton Friedman as socialist or anti-American. This is not because they have any idea who Milton Friedman was, which most do not, but because they are extremely conservative on almost every question. This no doubt is due in part to the self-selection built into the commissioning process, to the students' disproportionately rural and southern upbringing, and to the socialization process in the officer corps.

Professors frequently avoid assigning controversial articles on sensitive topics because the resulting hostile student reaction would be counter-productive and might produce evaluations that punish all the faculty. For example, the faculty has learned that it cannot assign readings from two of the most distinguished historians who have written on air power in World War II because of student reactions. Popular works of amateurs or memoirs of senior generals, on the other hand, find enthusiastic receptions if they follow the party line of the air power zealots. While many professors try to limit the introduction of controversial subjects into the discussion, such is not the case with all the colonels. Some of the latter in effect egg the students on by playing videos of Fox News commentators while others make conscious efforts to slant the discussion in an ultra-conservative direction.

The Washington AOR

An "area of responsibility" (AOR) is the military term for overseas regions where the military fights wars, drops bombs, and conducts campaigns of various kinds. Recently, the Air Force leadership began to use the term "Washington AOR" to refer to Congress and the Washington-area media. While some of this refers to inter-service rivalry, the term means much more than that.

Much of the Air Force's officer corps is openly hostile to what it regards as its liberal enemies in Congress. Such officers regard Congress, its staffs, and the media as enemies to be overcome or circumvented. Many officers are openly hostile to the national media except for the politicized conservative platforms. Much of what passes for "public affairs" is quite legitimate, but much else is thinly-veiled propaganda aimed at domestic audiences.

A Godly Place

Much has been written about efforts of evangelical Christians in the officer corps to use the military to pursue their fundamentalist religious agenda. The Air War College is no exception. Over the period 1991 to 2009 there have been numerous violations of proper policies on religion. For a number of years, attendance at the Thursday morning bible studies session, held in the commander's conference room, was nearly a prerequisite to holding a senior position in the Air War College. While interviewing for a job here, this author was advised to avoid any discussion of religion with a senior official at Air University headquarters. This person, prominent in a large Methodist church in Montgomery, frequently queried job seekers about their religious views during his portion of the hiring process.

One commandant openly pursued his evangelical agenda by sponsoring a religious-based marriage counseling service. Its purpose, he said, was to "equip military couples with god's plan for marriage and family." The same commandant later campaigned for Congress on an ultra-conservative platform that included the assertion that the United States was "doing God's work" in Iraq. Aggressive evangelical students, particularly chaplains, on occasion ask visiting senior officers what role their "faith" played in their career success or their leadership styles. In only one case did this author hear a general reply that faith was a private matter that need not be discussed in such a forum.

Other examples abound. A particularly zealous evangelical colonel tried to coerce a Jewish lieutenant colonel who worked for him into attending Christian church services. Many Air University and Air War College functions begin with openly Christian prayers. Some aggressive students insist on praying at official functions even when entirely inappropriate—for instance at a public dinner in a restaurant in the Czech Republic. Many emails at Air University close with religious salutations, while greetings at the base gate and in phone conversations frequently end with the closing "have a blessed day." At one time, visitors to both the commandant's office and to the Staff Judge Advocate School at Maxwell encountered the well-known but bogus painting of George Washington kneeling before a Christian Cross at Valley Forge. Air University, the Air War College, and even the Air Force at large are very godly places. Those who would prosper in these institutions should keep that in mind. This author has never seen an officer object to or try to correct any of this.

At Last: Pay and Benefits

The main reason for young academics to work a place like the Air War College is that the pay and benefits are better than unemployment or a long period of part-time or one-year appointments. The Air University offers its new professors higher starting pay (80,000 to 120,000 dollars) than does almost any university. The benefit package for federal employees is much less favorable than it was prior to 1983, and pales in comparison to that of the military, but is superior to many plans in other kinds of employment. Some faculty come to the Air War College or remain there because they enjoy the school's focus on national security or because of the availability of funds for short-term research trips and professional development. Still, very few tenured academics are willing to forego the academic life for the pay and benefits of Maxwell Air Force Base and the city of Montgomery, Alabama. They stay where they are, and for good reason.

Professors who wish to leave the Air War College to seek better school and cultural environments for their children must take positions at other military schools or join other government agencies. Their traditional academic careers end when they are hired at Maxwell. In the last 20 years, only a single civilian faculty member has been able to return to academe (at Alabama State University in Montgomery), although many have tried.

Chapter 11

No "Holidays from History": Adult Learning, Professional Military Education, and Teaching History[1]

Bradley L. Carter

> 11 September 2001 … abruptly ended the false peace of the 1990s as well as our "holiday from history".
>
> Secretary of Defense Robert Gates, November, 2007[2]

The Command and General Staff College

Imagine a classroom of no more than 18 students, aged 30 to 40, all with bachelor's degrees, most with graduate degrees, and all effectively on full scholarship, paying nothing for tuition, fees, or books. Imagine leading a weekly graduate seminar in history with an array of students whose professional backgrounds include medicine, nursing, ordained ministry, law, nuclear physics, engineering, and experimental aviation. Some of these students have taught at the nation's most competitive colleges; all have traveled extensively. Imagine a setting where not only are students typically well-read, absences rare, discipline problems all but non-existent, where seminar discussions are thoughtful, lively, engaged, relevant, candid and provocative.

Moreover, imagine a campus setting where not only are the students well-paid and cared for, but so, too, is everyone else. Promotions are merit-based, and pay is a matter of public record and surprisingly egalitarian. In fact, on "campus" the gap between the lowest and highest-paid institutional employees is less than a factor a ten—that is the highest-paid person makes less than ten times more than

1 The views expressed here are the author's and do not reflect those of the Department of Defense, The Department of the Army or its elements. Thanks to Colonel (retired) Thomas McShane, Supervisory Professor and Team Chief, US Army Command and General Staff College, Fort Gordon (Georgia) Campus, Mr. Steve Rauch, Historian, US Army Signal Center, Fort Gordon, Georgia, and Colonel (retired) Mark Montesclaros, Associate professor of Joint, Interagency, and Multinational Operations, CGSC, Fort Gordon, for their invaluable assistance with this chapter.

2 Gates 2008: 5.

the lowest.[3] All students and "campus employees" also receive a generous benefits package that includes a housing allowance, all-but-guaranteed annual raises and extensive medical care for themselves and their families. Imagine as well a state-run medical system providing care for all without any out-of-pocket expenses for students, employees, and their families.

The teaching setting described is neither fictional nor located in some progressive Scandinavian utopia. Rather it is an accurate—if highly selective—description of the United States Army Command and General Staff College (CGSC) and the Command and General Staff Officer's Course (CGSOC).[4] Based at Fort Leavenworth, Kansas, the College includes several schools offering graduate-level education through a range of programs including the CGSOC. The mission of the college's largest school—the Command and General Staff School which oversees the CGSOC—is to educate and train "intermediate level Army officers, International Officers, Sister Service Officers and Interagency leaders prepared to operate in full spectrum Army, joint, interagency, and multinational environments as field-grade commanders and staff officers" (United States Army Combined 2009).

Since January 2006, I have been privileged to work as an Associate Professor in the Command and General Staff School's Department of Military History. For the past three years I have taught graduate-level courses in military history (and other subjects) to Army officers, mostly majors mid-point in a 20-year career.[5] This array of students includes not only the uniformed professions mentioned above, but also other medical specialists, social workers, "foreign area officers," Special Forces officers, and numerous West Point faculty members, in addition

3 In 2009 the basic monthly pay for a US Army private (E-2) with less than one year's service was $1,568.70; basic monthly pay for general officers is capped at $14,750.10. Compare that differential to that on many college and university campuses, where pay for senior executives and head coaches can be exponentially greater than that of the school's lowest paid employees.

4 The Command and General Staff College was founded at Fort Leavenworth in 1881. The College currently consists of the Command and General Staff School, the Department of Distance Education, the School of Advanced Military Studies, the School for Command Preparation, and the Army Management Staff College. All Army majors are required to complete the four month CGSOC through the CGSC in order to be eligible for promotion to lieutenant colonel. The CGSOC is offered in-residence as part of a ten month program at Fort Leavenworth, and—since 2005—at three satellite campuses located at Fort Belvoir, Virginia, Fort Gordon, Georgia, and Fort Lee, Virginia.

5 My primary teaching assignment is a twelve-lesson class, "History 100: The Rise of the Western Way of War." I typically teach four "Staff Groups" (individual classes of up to 18 students) in two hour seminar sessions once a week, three times a year. I also share responsibility with other faculty to teach blocks in Critical Thinking, Media, Culture, Civil-Military Relations, Self-development/Leadership, and Writing as well as in additional mentoring, administrative and leadership duties as a "Staff Group Advisor" or to an individual class.

to the more-expected mix of armor, infantry, artillery and logistics officers. It is with this surprisingly diverse student population that I teach an Army-directed curriculum that includes a "mandatory humanities elective."[6]

Faculty Training and Adult Learning

This job has required negotiating a new professional role within an organization that is simultaneously a graduate-level institution and part of the United States Army. At times it has felt like a sociological experiment in which I've been a participant-observer. I came to this job with an interdisciplinary PhD in American Studies as a "pure civilian"—that is, with no prior military service. Most of my faculty colleagues are recently retired colonels or lieutenant colonels. They've spent the past 20 years as professional military officers. In contrast, I've spent the past two decades as a seminary student, then administrative dean at a divinity school, then as a doctoral student and instructor at a large civilian university.

Teaching history to Army officers differs little from teaching seminary or undergraduate students. What is different is the teaching environment. Adapting to this new institutional setting—and its prescribed teaching philosophy and military culture—has been the greater professional challenge. This chapter is informed by that cross-cultural experience of relocating from a civilian to a military school. It offers a theoretical re-conceptualization of adult learning in a professional military setting by challenging the institutional and pedagogical assumptions of the "adult learning model" that CGSC faculty are expected to use. It also draws upon the experience of negotiating a teaching role in this new setting to discuss the teaching milieu itself, students, colleagues, and culture.

At CGSC I function at the center of an on-going debate over "how best to advance professional education in the US military" (Petraeus 2007).[7] The US Army is no stranger to anti-intellectualism.[8] However, the "war-fighters" that comprise the twenty-first-century US Army include an officer corps that is professionalized and more highly educated than at any period in its history. Rather, the fault lines today tend to be over "the relationship between the Services and the civilian academy" (Petraeus 2007). CGSC attempts to bridge that gap with an increasingly civilianized faculty who—in the CGSOC common core curriculum—are charged with preparing "field grade officers with a warrior ethos and war fighting focus for

6 In addition to Military history, the CGSOC curriculum common to the satellites and Fort Leavenworth is taught by faculty representing the Department of Logistics and Resource Operations, Joint and Multinational Operations, the Center for Army Tactics, and the Center for Army Leadership. H100 really is the curriculum's closest equivalent to a humanities elective—albeit a required one.

7 For an opposing view see Peters 2007.

8 For a fuller discussion of anti-intellectualism and the US Army see Matthews 2002a and 2002b.

leadership positions … [and] conducting full spectrum operations" (United States Army Combined 2009).[9]

An often-repeated maxim across the college is that the CGSC curriculum aims at "teaching students *how* to think, not *what* to think." Moreover, the school attempts to distinguish teaching—and attending development of critical reasoning, independent thought, and higher learning—from the more ubiquitous military practice of training. As another popular maxim goes, you "train for certainty, but educate for uncertainty." Drill, in every sense of the word, is surprisingly absent from the CGSC curriculum, classroom and campuses.

Yet, before "taking the podium"—military-speak for teaching a classroom of actual students—new faculty, civilian and military alike, are required to undergo a two-week training program focused on both how and what to teach. This "Faculty Development Program," like many things in military culture, is better known by its acronym, "FDP." The first week of FDP (nomenclatured as "FDP1," naturally) serves as an introduction to the "Adult Learning Model," the pedagogical model central to the CGSC curriculum.[10]

Since all CGSC students are (typically age 30-ish) adults, the "Adult Learning Model" makes sense. It is predicated upon the belief that adult learners are self-motivated and choose to learn what is most personally beneficial. This model taught in FDP1 is designed to counter models based upon repetition, drill, and rote memorization that are hallmarks of military training. It is well-suited for classrooms of highly-motivated adult students engaging in the experiential learning CGSC provides.

The existence and dissemination of a clearly defined teaching philosophy/model mark a notable departure from my previous teaching in civilian settings. As a new graduate teaching assistant at the University of Kansas in the mid-1990s, I attended a half day's worth of workshops for new teachers, focused primarily on how to spark classroom conversation from laconic (if not hung-over) discussion sections, coupled with a day's worth of departmental meetings. That was it. Like most of my colleagues with backgrounds in the liberal arts, educational theory wasn't something we had much knowledge of (nor, to be candid, much respect for, considering it the bailiwick of departments or schools of education). Training was mostly on the job, assisted by peers, colleagues, and some gifted mentors. By hook or crook we learned how to teach through an apprenticeship of doing.

9 The effort to leverage civilian models of higher education for "professional military education" (PME) is not without its own tensions and challenges—especially at CGSC—as discussed below. Of CGSC's nearly 400 full-time faculty, 85 percent are retired military, mostly retired field-grade officer (Majors, Lieutenant Colonels, and Colonels). Seventy-five percent of faculty have at least a Master's degree; 25 percent have doctorates, with another 10 percent enrolled in doctoral programs. CGSC is accredited by the North Central Association and offers students the option of earning a Master's degree (even though most elect not to undertake the additional work required by the school's degree options).

10 This model is largely drawn from Houle 1961.

By contrast, CGSC promotes the adult-learning pedagogical model in an intentional, top-down manner. The week-long FDP1 course indoctrinates all new faculty in that model, training them to conduct classes that include—in sequence—a "concrete experience," time to "publish and process" material and opportunities to "generalize new information" (United States Army Combined 2009). New faculty are required to demonstrate their ability to teach a lesson, by those numbers, in order to be certified platform-worthy.

Since all lessons in the CGSC curriculum are written in accordance with that model, a week's training helps new faculty break the code on the lessons they're given. However, the danger for new faculty, especially those whose traditional military backgrounds render them most comfortable with a rigid, by-the-numbers training-focused approach to classroom instruction, is an overly rigid adherence to the teaching model. The question taught in FDP1, and found in many lesson plans—"now that you know this, how might this apply to your military career?"—can quickly become meaningless boilerplate.

At its best, though, a mandatory, school-wide practicum on learning theory and classroom instruction is invaluable. It promotes collegiality across departmental boundaries while ensuring a degree of standardization among diverse teaching departments. It introduces new faculty with varied backgrounds to a culturally specific way of teaching. And it facilitates new faculty in becoming better teachers in an environment where education and critical thought really are valued more than rote drill and mission- or task-oriented training. The adult learning model embraced by CGSC is at its heart an adaptation of classical Socratic methods. Its privileging of experiential learning within a peer-led seminar setting can—and in my experience often does—create opportunities for remarkable teaching moments. The Socratic image of teaching as midwifery, of helping give birth to that which on some level was already known, can be fully realized in CGSC classrooms filled with motivated adult learners.

There exists, however, a significant disconnect between the CGSC's learning model and reality. The free-agent, volitionally present and self-motivated student presumed by the adult learning model is not exactly the CGSC Army (or sister-service) officer who is at best "voluntold" to attend either the ten-month course at Fort Leavenworth, Kansas, or one of the four-month satellite courses offered at Forts Belvoir and Lee, Virginia and Fort Gordon, Georgia. While I joke with my students that their military history course (consisting of 12 two-hour weekly seminars) is their humanities elective, since they are literally ordered to attend, there's not much that's really elective about it. There are "no holidays from history," at least in a curricular sense. At the same time, the notions of motivation and voluntary presence central to adult learning are nuanced. In a literal sense all CGSC students are volunteers, having volunteered for service in an "All-Volunteer" military. While they "have" to be in the CGSC Officer's Course—including the military history class—in many ways these students are no different from the undergraduate students who "chose" to take the Humanities and Western Civilization course I taught that fulfilled a graduation requirement at the University of Kansas.[11]

Despite the cultural and contextual differences, teaching military history with middle-aged military officers is not that different from teaching "Western Civ" to traditionally-aged undergraduates. The liberal arts are the liberal arts. Liberal learning is liberal learning. Students show up (or don't show up) for a range of often over-determined reasons. Good discussion, good thinking, good writing, and good learning are good in whichever setting. Most of all, the pedagogical challenge is the same: motivating students to critically engage the material in ways that can make a difference in their lives, and helping them understand both the material and those connections to their larger lives.

That challenge can be mapped out in a way that also reconceptualizes the CGSC teaching model:

Table 11.1 CGSC teaching model

	Training	Education
Internally-Driven Motivation		(CGSC Ideal)
Externally-Driven Motivation	(most military instruction)	(CGSC Actual)

The distinction between training and education is represented by the two vertical columns. A distinction between internal and external motivation is represented by the two columns that align horizontally. In this model, the majority of military instruction occurs in the lower left quadrant—it is task or skill-directed training that is externally mandated and driven. All service members receive initial military *training* of this type. The difference between the mandatory military training represented in that quadrant and what a student enrolled in a vocational training program might be one of motivation and direction. The latter could be represented in the upper left quadrant.

The learning model that informs CGSC instruction, however, presumes a student population residing in the upper right hand quadrant—internally driven students wanting education; voluntary learners interested not only in whats and how-tos, but also the critical thinking of whys, why nots and what ifs? The reality is that on an average day a CGSC classroom (while conceivably embodying elements of all four quadrants) most typically occupies the lower right hand quadrant. "Education for uncertainty" rather than "training for certainty" is likely to occur, but with students who are there because they have to be instead of want to be. In that way a CGSC classroom resembles one of undergraduates taking a

11 Since completion of CGSOC is a prerequisite for promotion from major to lieutenant colonel, the motivation of CGSC students aligns even more closely with undergraduates taking prerequisites toward a tangible goal.

class to fulfill a liberal arts distribution requirement for graduation. For a teacher both classes present the same challenge—motivating students to become active learners, using the tricks of the trade to get them to dig critically into the material in ways that add meaning to their lives. In both cases, too, it is ultimately up to the student whether they learn.

Although CGSC's military history courses are the closest the curriculum comes to the traditional liberal arts, the professional mission of the school necessitates a practical component. In that way, CGSC's mission is similar to those of the theological seminaries of my earlier academic experience—both are graduate professional schools educating (and in some cases training) practitioners in and for a specific vocational field. That focus makes the task of transforming externally motivated into internally motivated learners all the easier. In fact, one of the questions asked in my interview for the CGSC job centered on how I would approach the challenge of helping students connect the history curriculum to their military careers. My answer was something along the lines of, if I could help 20-something non-philosophy majors find meaning in Aristotle, then how much tougher could it be to help military officers find something useful in military history?

The idea of a "usable past" is central to CGSC's Department of Military History's curriculum.[12] The second phase of FDP—FDP2—involves a week-long orientation to that curriculum. At first blush this pragmatic focus would seem to distinguish CGSC's history department from a traditional university setting. The curriculum is aimed at the professional development of practitioners. Yet the content would be familiar to any military history graduate seminar: "Revolutions in Military Affairs," Gustavus Adolphus, Frederick the Great, Napoleon, Jomini and Clausewitz, industrialization and the American Civil War and World War I, the inter-war period and various aspects of World War I, and so on. The lessons tend to be Euro- and US-centric, and—befitting students who are making a professional transition from being hands-on company grade officers to field grade officers—privilege operational and strategic-level analysis over the tactical level.

This pragmatic approach to history is a site of some tension. The notion of a "usable past" is not and has not been universally held. Some students simply are not inclined to find anything useful in nineteenth-century military theory or operations. Other concerns are methodological. The Department of Military History uses a thematically and chronologically arranged curriculum, supported by assigned readings from a mixture of primary and secondary sources. This seminar/survey approach is sometimes at odds with others within the Army who prefer a case-study method. A post-September 11 keenness for mining a usable past—especially lessons to be learned from earlier "counter-insurgency" fights, is evident in the historical vignettes that pepper the 2006 Field Manual 3-24,

12 See Luvaas 1982. Luvaas' article is the first reading assigned CGSOC students in the History 100 course.

Counterinsurgency. Other recent Army publications similarly highlight historical examples to illustrate current doctrine.[13]

Still, student resistance to the curriculum remains. A notable minority of students equate a usable past with recency. The current basic history course curriculum—the one all CGSOC students receive at Fort Leavenworth as well as the three satellites—ends with World War I. Some students would prefer a history course based solely on events since 2001. The departmental mantra, however, is "recency does not equal relevancy." Again, the pedagogical task is to help students connect the dots. Here my background in teaching the humanities is especially applicable: I believe liberal learning is always useful (even if pragmatism isn't the only value that should be attached to education). Discussing the Clausewitzian concepts of "friction" and "fog" in battle are timeless, relevant, and applicable. So, too, have been the World War I poems of Wilfred Owen and Siegfried Sassoon (or even the Dropkick Murphys' recording of "The Green Fields of France" or the Pogues' cover of "And the Band Played Waltzing Matilda").

The experiential element of the adult learning model actually makes discovering a usable past easier. The overwhelming majority of CGSC students have had one or more deployments to Iraq or Afghanistan.[14] Discussing military history—not only planning and operations, but especially the human "face of battle" and the painful realities of war—with students who have lived that reality is perhaps the most remarkable aspect of teaching at CGSC. It is the poor student (or teacher) who cannot find relevance and a usable past in that classroom setting. Again, my task is the midwifery that enables experience-based knowledge to give birth to usable education.

Moreover, a usable past as understood in the CGSC curriculum is not based upon a simple, deterministic understanding of history. History does not exactly repeat itself, nor is its study directly predictive. Content matters, but history becomes an opportunity to apply critical thinking to the past. Its usefulness emerges in learning to ask the right questions, in gaining perspective, and in understanding human nature, culture and processes. It serves as a tool enabling new ways of thinking. It is truly about how and not what to think. This approach in turns supports the Army's institutional goals of developing officers who are capable of critical thinking and adaptive leadership.

13 The historically-informed nature of the counterinsurgency field manual is no accident. It was largely written by a brain-trust of current and retired Army officers with advanced civilian degrees (Peter Mansoor, Conrad Crane, John Nagel) under the direction of then Combined Arms Commander (which oversees CGSC), then-Lieutenant General David Petraeus. General Petraeus, who earned a PhD in international relations at Princeton, has been a vocal advocate for greater *civilian* graduate schooling for military officers. See Petraeus 2007.

14 Not to mention other war areas students have served in, including the very occasional (older) student who is a Vietnam War veteran.

Using a prescribed curriculum with mandated readings was nothing new. In the Humanities and Western Civilization Program at the University of Kansas, where I taught for eight years, the curriculum was similarly dictated. And, earlier in my career, my seminary work was within a religious tradition that used a prescribed "Lectionary" of weekly scripture readings. All three settings present the shared challenge of adapting pre-assigned material to a specific audience. CGSC's Department of Military History encourages innovative, creative use of the material, granting latitude to individual teachers in tailoring the assigned lessons to their settings. As I tell my students, the challenge—the fun even—comes not in trying to color off the page, but in finding new ways to color *on* the page, even when that means coloring well outside the lines.

What was new was adapting to the prescribed lesson *plans*, the accompanying "Learning Objectives," and the ubiquitous Power-Point slide packet. A top-down curriculum with ready-made slides risks a canned, "death-by-Power-Point" classroom. That has not been my experience. On good days, what happens in a CGSOC seminar rivals the best of any undergraduate honors or graduate seminar discussion. Ultimately, curricular and pedagogical differences between civilian and military teaching settings are less significant than cultural and environmental ones. Adapting to CGSC students, the institution and its expectations, and working as part of team of military colleagues has been more challenging, as well as rewarding, than negotiating FDP1 and 2 or learning a new curriculum.

Student Culture at CGSC

Overall, Command and General Staff College students are among the brightest students I have taught. Compared to undergraduates in the liberal arts, other professional graduate seminary students, and even honors seminar students, my CGSC students are better-read and consistently engaging in thoughtful, critical seminar discussion. When discussions veer toward subjects like torture, US military involvement in Iraq and Afghanistan, "Don't Ask, Don't Tell," and even topics like the US war in Vietnam or the history of US imperialism, the range and depth of student discussion exceeds any I've encountered in other academic settings. Candor flourishes in a closed-door classroom with a policy of "non-attribution."[15] These thoughtful professionals bear little resemblance to stereotypes that depict military officers as ignorant, trigger-happy knuckle-draggers. Nor are students

15 An example of such candor that received widespread attention within and outside of the military is Yingling 2007. Yingling's barbed critique is now assigned reading in the CGSOC curriculum. Controversial at the time and still, it exemplifies what the best CSCG seminar discussions aspire to. Since it liberally uses historical examples from the Department of Military History's curriculum that Yingling had as a CGSC student, there is speculation among history faculty that Yingling's article is actually a reworked version of the final argumentative essay required by the department.

unquestioning martinets or automatons. In fact, students (to use their own argot) are quick to "throw the bullshit flag" on peers who uncritically parrot back the "right answer," accusing them of "drinking the kool-aid."

Creative, critical thinking is not without its limitations, however. CGSC students seldom stray into radically unconventional territory.[16] They are sharp and engaged, but they are not creating a culture of intellectual bohemianism. After all, this is professional education for an inherently conservative profession. At the same time, the conservatism exhibited by CGSC students tends to be more pragmatic than ideological. A "conserving" impulse makes sense given the nature and mission of the military. Pragmatism pervades this culture whose business is violence and death. Get too creative, too funky, stray too far, too fast from the received wisdom and way of doing things and people end up getting killed. Get too wound up around a particular political or religious ideology—as students often point out—and even more people get killed. And these students are among the people doing the killing and dying. If anything, the political conservatism of many students is much less closely held and stridently championed than what I've seen among some undergraduate College Republicans.[17]

The pervasiveness of evangelical Christianity within the US military is also evident among CGSC students.[18] Early morning Bible study groups are common. Evangelical Protestant materials—for example, copies of Rick Warren's best-selling *The Purpose-Driven Life*, or donated copies of the New Testament or other devotional literature—appear in student break or work areas with a frequency unimaginable in a public university setting. Yet, contrary to some more alarmist claims, at least at CGSC, evangelical Christianity enjoys neither a hegemonic presence nor poses a pernicious threat. Theological discussions (moreover disagreements) are non-existent in the classroom. Any Evangelical presence is balanced by equal numbers of students who are Roman Catholic or identify with no religious group. Smaller numbers are from "mainline" Protestant or Jewish traditions. Folks tend to be mutually tolerant, if not respectful, of religious differences. Students would rather "cooperate and graduate" then pick sectarian fights. Pragmatic professionalism prevails.

Countering student cynicism is a teaching challenge greater than negotiating ideological minefields. In their ten-plus years of military service, most students have been deployed multiple times. For some, an experiential educational model

16 The culture also carries with it other discursive boundaries that affect teaching. For example, discussions of pacifism—usually in the context of a reaction to the horrors of World War I—are awkward and rare. Students can identify with the poetry of Sassoon and Owen, but given the profession for which they have volunteered, considerations of nonviolence as a philosophical construct are too countercultural to be easily engaged in.

17 Kohn 2001 and 2008 has written extensively on the reasons for and implications of the trend toward a Republican Party-identified officer corps in the context of a growing civilian-military cultural gap.

18 The most comprehensive study of this growing phenomenon remains Loveland 1997.

emphasizing critical thinking within a pragmatic culture can turn healthy skepticism into outright cynicism. While overtly nihilistic cynicism is rarely expressed, I have noticed it among a growing—but still tiny—handful of students, almost all back from multiple deployments. I believe this cynicism is often a symptom of larger readjustment issues. Post-Traumatic Stress Syndrome, and other post-deployment problems—including soldier suicides—are matters of real concern for the Army and CGSC.

At the same time, these challenges also create teaching moments unlike any in a civilian classroom. For example, using Randall Jarrell's poem, "The Death of the Ball Turret Gunner," in a class of combat veterans is different from using it with undergraduates. As a "pure civilian" who has not been to war, I failed to anticipate the poem's potential classroom effect. The first time I used it in a CGSC seminar the room's energy immediately changed. I learned that the poem's grimly evocative final line, "When I died they washed me out of the turret with a hose" echoed the experience of many of my students. At first reluctantly, they told their stories of hosing out military vehicles, of washing out the remains of comrades and friends. As the discussion continued I learned that the preferred method is to ask people from a neighboring unit to do the job, with the expectation you would return the favor for them. That discussion illustrated for me the value of experiential adult learning, while also reminding me of a significant experience gap between me and my students.

When I began this job I worried more about fitting into a professional military setting than about anything directly tied to curriculum or teaching. Reflecting the larger officer corps, CGSC classes tend to be 80 percent male. I knew military officers defied the worst brainless Rambo-type warrior caricatures, but was not surprised by the highly competitive, type-A nature of the culture. In the Army, "guys" really is a gender-inclusive term that includes "females." Moreover, discussions both in and out of class often use—and sometimes revel in—an amazingly profane vocabulary. Four years in a fraternity house have provided invaluable preparation for adapting to an academic context that is often more Army than schoolhouse, reflecting a culture that remains macho, sometimes crude, and prone to pranks and "ass-grabbery," as well as jokes, even if the joking is often of the "that's not what your wife said last night" variety.[19]

In making this adjustment I have discovered that the Army is most of all a mission-oriented culture that values competency and team-work. My anxieties about fitting in were overcome not so much by getting in touch with my inner

19 A frat house/locker room type of casual homophobia remains a vernacular element of this culture. Malicious homophobia is rare; rarer still is outright sexism. In my experience the jocularity is good-natured and usually well-intentioned, however politically incorrect or insensitive it might be. Personal attacks, overtly hostile remarks, or anything that creates a hostile environment are officially proscribed and very few. On the handful of occasions when a student has crossed that line in class, I have experienced no problem calling them out for their remarks and having that line respected.

frat-boy as they were by exhibiting confidence and demonstrating competency. Knowing your stuff, giving a damn, and making an effort to be a team player all matter at CGSC. In the Army competency and commitment lead to acceptance. Conversely, screw any of that up and you lose effectiveness, civilian and uniformed military faculty alike. In many ways, though, that experience is little different from teaching civilian students. Students want to know if a teacher knows their stuff, cares about them, and is fair.

The most significant adjustment has been to the mission and team orientation of the military. The Army is a highly competitive culture that functions with an exceptionally high degree of cooperative selflessness. Team-building and teamwork are taken seriously. This aspect of military culture is perhaps most at odds with the traditional academic world that prizes individual scholarship. At its worst, that world becomes a solipsistic universe of isolated academics. Whereas many academic departments seem to be a collection of disparate scholars who are "all about me," at CGSC there are very few "about me" moments. A welcome manifestation of this is faculty meetings largely acrimony-free and seldom lasting more than half an hour.

Academic and Military Cultures

I have experienced an unusually high level of collegiality among my CGSOC colleagues. While bureaucratic and interpersonal politics are unavoidable, they are less frequent and less petty in a military setting than they are in the highly-balkanized, decentralized structures of civilian institutions. There is little tolerance for grandstanding, prima-donnas, and "chickenshit." Part of that is a result of a culture that is exceptionally task-focused. It is also a by-product of a curriculum that utilizes teaching teams. Sharing an office with two or three peers—itself a significant adjustment from a private office—also enhances collegiality, especially when those office mates are from different academic departments.[20]

Although there is an institutional effort to recruit faculty from civilian academia, the majority of faculty are retired military officers. These retired colonels and lieutenant colonels tend to bring the military culture they knew on active duty with them to their new roles as CGSC faculty. As such, faculty pull together and take on additional responsibilities beyond teaching-related tasks. Especially at the satellite campuses, faculty members serve as property book custodians, credit card holders (for purchasing supplies through the government system), librarians, and information technology coordinators; moreover, when textbooks arrive, the entire team (and students who happen to be nearby) helps to unload books off the truck

20 At the main CGSC campus at Fort Leavenworth, faculty share offices, too. However, they typically share offices with members of their respective departments. The sharing of offices across departmental boundaries tends to occur only at the three satellite campuses.

(McShane 2008: 28). Hauling boxes of books, cleaning classrooms, moving and assembling furniture are all part of a day's work as a CGSC professor—tasks I sometimes have trouble imagining civilian colleagues performing at all, much less with relative enthusiasm.

These are also colleagues in a culture for whom integrity and "Army Values" matter, too. People presumably say what they mean. Duplicity and the pettiest of personal attacks are rare. And your word really is your bond. I was reminded of this early in the job when I told my boss that I was going to mail something later that afternoon. The following day I mentioned something about having dropped the item in question in that *morning's* mail. Although the mailing was business-related, it wasn't especially time-critical. Still, he let me know that I had indicated I was taking care of the matter yesterday, not today. His admonition was a reminder of the cultural divide I had crossed. In a profession with lives at stake, if you say you're going to do something at a specific time, it is understood that you will. Failing to do so, even without malice or real consequence, is not the Army way.

For all of its exceptional collegiality, this new professional setting is not immune to bureaucratic absurdities. This is the Army after all. What's different is its scale and nature. Being located on a larger military post means our faculty office computers are managed by the installation's "Department of Information Management" (DOIM). Internet access is filtered according to the concerns of a military activity (in this case, Fort Gordon) whose primary mission is *not* graduate-level education. As a result, the same web filters designed to keep 20 year old soldiers from accessing inappropriate material or goofing off at work are applied to faculty. Those filters have been successfully broadened for faculty, but only after documenting how the filters designed to block out hate groups also blocked web pages that mentioned the Holocaust, and how anti-violence filters blocked anything pertaining to Civil War weaponry—sites of obvious use to someone teaching military history![21]

The bureaucratic demands of being a Department of the Army employee (albeit with an academic title) have required other adjustments. CGSC faculty serve under renewable, multi-year contracts. There is no tenure. Moreover, faculty submit bi-weekly time sheets and are expected to put in regular 40-hour work weeks. "Duty days" consist of eight hours "on platform" or in your (shared) office, twelve months a year. Absences must be pre-arranged and/or approved, subject to policies addressing sick time or annual leave. The traditional academic rhythm of working at home, extended periods of non-teaching time available for research, and the exceptional level of flexibility and autonomy that the (civilian) academic profession affords remain alien to the CGSC's military culture. Working outside

21 Sometimes the filter's logic is less discernible. For a long time both the website for the public radio program, Prairie Home Companion, and Susan Sontag's homepage were blocked. The struggle over Internet access continues. CGSC students are now required to post a comment on a "blog" as part of a "Strategic Communications" requirement. However, until very recently, most "blog" pages were blocked by various installation DOIMs.

of the prescribed boundaries of a 9-5 work day (more like "0700-1600") requires careful negotiation with superiors. Such arrangements are possible, but they threaten the dominant culture in other ways. As retired officers, some faculty colleagues are unaccustomed to the independent and idiosyncratic work patterns of academia. Most have experience as trainers in the military; some have experience teaching in civilian institutions. Still, they remain primarily acculturated to a military rather than an academic workplace. Being present, hanging out in an office for a regular duty day—productive or not—constitutes "working." However, absence from an assigned office—even for class preparation, grading, individual research or writing—risks being perceived as *not* working, as not contributing to the team and the mission, even when the mission is education. Ironically, sitting at a desk *reading* runs a greater risk of being perceived as non-work than sitting at that same desk surfing (non-blocked) websites, chit-chatting with colleagues, or watching the television placed in each faculty office.[22]

At the same time, CGSC faculty are compensated better than most of their civilian peers. Not only is the pay generous, so too is the benefits package. As employees of the Department of Defense, faculty are also assured of annual cost-of-living raises. Since most faculty are also retired officers, their faculty pay is in addition to their retirement pay and benefits.

Other professional adjustments illuminate the institutional differences between a civilian and a military academic setting. Student absences are largely a non-issue. Theoretically an absence could constitute being "Absent Without Leave" and subject the student to discipline under the Uniform Code of Military Justice (UCMJ). Academic misconduct is also covered by the UCMJ. Plagiarism is significantly less prevalent among CGSC students. When plagiarism is suspected, however, the procedures involved are significantly more serious than in a civilian setting. If a written assignment aroused my suspicion at the university—or even the seminary—I would call a student in for an informal discussion. At CGSC, faculty are required to have an active-duty officer of greater rank than the student involved present to read the student their legal rights under Article 15-6 of the UCMJ *prior to* that conversation. Depending on the circumstances and severity, a formal investigative board may be formed and—theoretically at least—a student could face the prospect of doing the "long course": confinement in the Disciplinary Barracks, or military prison, at Fort Leavenworth. Here plagiarism is literally a crime.

A recent article on "The Selective Use of History in the Development of American Counterinsurgency Doctrine" serves as a final reminder of the most significant difference between civilian and military academic settings. In his *Army History* article, Gian Gentile (2009) challenges the historical and theoretical underpinnings of the Army's new Counterinsurgency Field Manual. Head of West Point's military history program, Gentile furthers an on-going academic

22 CGSC does offer block time off for research and writing. Sabbaticals *per se* do not exist, but these times are available for faculty on a competitive basis.

discussion. As an active-duty colonel at an institution similarly charged with educating the next generation's officer corps, Gentile is engaged in something beyond an academic debate. Whether at the United States Military Academy or the Command and General Staff College, debates are never simply academic. They have real world, life-and-death consequences. This is professional education for a serious profession. "Ivory towers" and "front-lines" are connected in a way that grounds even the most theoretical of debates. Both West Point's and CGSC's History Departments therefore reflect their institution's larger educational missions—the usable past we teach should be usable to students engaged in the profession of arms.

That final point highlights an unresolved area of professional concern. However successfully I have negotiated this new role, there remains a "larger gap between civilian and military spheres in American political culture, especially within the academy" (Kusiak 2008: 75). I have maintained professional and collegial relations with my civilian peers during this transition. At the same time, I wonder if working for CGSC has brought upon me the "stench of the Army." Coming out of an academic background (mainline Protestant theological education and American Studies) that is perhaps even more ideologically skewed liberal than the military is in the opposite direction, I am well aware of a sometimes reflexive anti-military bias. Some of that bias is well-founded and reflects progressive values I personally share. Some of it is also based on nothing more than ignorance or outright prejudice. The most recent "academic backlash" over "coziness" with the military has been loudest in social science circles—especially anthropology—focused on the military's use of civilian scholars on "human terrain teams" in Iraq and Afghanistan. To date, those tensions are less rancorous among historians. Part of that might be because the Departments of History at West Point and CGSC represent the two largest employers of PhD military historians in the country. Nevertheless, while my (and my colleagues') experience of teaching for the Army's Command and General Staff College is to the contrary, too many civilian academics still believe that the "military's culture of uniformity and discipline, important as they may be for an army, hardly aligns with those pedagogical values" of critical thinking and higher education (Kroll 2009).[23]

Conclusion

First-hand professional experience as an educator with the Army has shown me that critical thinking and quality education are possible. Moreover, I have found the civilian perspective I bring to be welcomed. The military seeks to be an adaptive, self-correcting culture. As an outsider I can offset a tendency toward group-think. My experience with graduate professional education and broad background in

23 Bridging that gap was the theme of April 2009 conference hosted by Harvard Divinity School, "Ivies and the Military—Toward Reconciliation." See Schaeffer 2009.

humanities teaching at a university has served me well. Not familiar with the larger Army's training mindset, I have less to unlearn and been able to help realize CGSC's desired adult-learning model of education. Most of all, the work is little different from teaching students in other schools. Helping spark student self-motivation—moving them from the lower right to the upper right quadrant of that grid—varies little from one teaching situation to the next. What is different, what has been most challenging, has been fitting into a new culture that is both academic and military.

As the Command and General Staff College continues to civilianize its faculty, more scholars will find themselves making similar adjustments. Civilian academics with no prior military service can be effective "force multipliers" for the school's educational mission. They can adapt and function effectively, benefitting from the experience as students benefit from them. At the same time, the institution will face increasing challenges fitting civilian faculty into a military culture. Institutional expectations for increased scholarly production pose potential conflicts with the school's traditional focus on teaching, not to mention an Army culture that is not always sure how to accommodate academics used to working autonomously at odd hours. A new fault line might yet emerge, reflecting one found elsewhere in higher education: will the school remain a teaching institution, more like a liberal arts college, or follow the path of a larger, research-oriented university? So far, CGSC is attempting to do both, while simultaneously remaining faithful to its primary mission of providing professional military education.

Bibliography

Ahern, D. and Shenk, R., eds. 1982. *Literature in the Education of the Military Professional*. Colorado Springs, CO: Department of English and Fine Arts, United States Air Force Academy.

American Anthropological Association. 2009. *Final Report on the Army's Human Terrain System Proof of Concept Program*. [Online]. Available at: http://www.aaanet.org/cmtes/commissions/CEAUSSIC/upload/CEAUSSIC_HTS_Final_Report.pdf [accessed: March 15, 2010].

Anderson, C. 2008. Wallflowers do well. *Navy Times* [Online January 21]. Available at: www.navytimes.com/community/opinion/navy-opinion-letters-080121/ [accessed: June 1, 2009].

"Army Regulation 870–5." 2007. *Military History: Responsibilities, Policies, and Procedures*. Washington D.C.: Headquarters Department of the Army.

Arnold, H. 2005. *Interview with Lieutenant Colonel Henry A. Arnold.* [Online: Operational Leadership Experiences in the Global War on Terrorism]. Available at: http://carl.army.mil/contentdm/home.htm [accessed: January 27, 2010].

Bain, K. 2004. *What the Best College Teachers Do*. Cambridge, MA: Harvard University Press.

Balibar, E. 1991. Citizen subject, in *Who Comes After the Subject?* edited by E. Cadava, P. Connor, and J.-L. Nancy. New York: Routledge, Chapman and Hall, 33-57.

Barber, R. and Parker, S. 2008. Evaluating Iraq's Provincial Reconstruction Teams while drawdown looms: A USIP Trip Report. *The United States Institute of Peace* [Online, December]. Available at: http://www.usip.org/resources/evaluating-iraq-s-provincial-reconstruction-teams-while-drawdown-looms-usip-trip-report [accessed: November 18, 2009].

Bercuson, D.J. 2009. Up from the ashes: the re-professionalization of the Canadian forces after the Somalia affair. *Canadian Military Journal* [Online], 9(3). Available at: http://www.journal.dnd.ca/ [accessed: November 1, 2009].

Boas, F., ed. 1944. *General Anthropology*. War Department Education Manual EM 226. New York: D.C. Heath.

Bok, D. 2001. Universities and the decline of civic responsibility. *Journal of College and Character* [Online], February. Available at: http://www.collegevalues.org/articles.cfm?a =1&id=570 [accessed June 1, 2009].

Braudy, L. 2003. *From Chivalry to Terrorism: War and the Changing Nature of Masculinity*. New York: Knopf.

Burke, C. 1992. Dames at sea. *The New Republic* 207 (8/9), 16-20.

Burke, C. 2004. *Camp All-American, Hanoi Jane, and the High-And-Tight: Gender, Folklore, and Changing Military Culture*. Boston: Beacon Press.

Butler, J. 1993. *Bodies that Matter: On the Discursive Limits of "Sex."* New York: Routledge.

Butler, J. 2005. Contagious word: paranoia and "homosexuality" in the military, in *Queer Theory*, edited by I. Morland and A. Willox. New York: Palgrave Macmillan 142-57.

Ciesinski, P. 2006. *Interview with Lieutenant Colonel Paul Ciesinski*. [Online: Operational Leadership Experiences in the Global War on Terrorism]. Available at: http://carl.army. mil/contentdm/home.htm [accessed: January 27, 2010].

Clay, S.E. 2007. *Iroquois Warriors in Iraq*. Fort Leavenworth: Combat Studies Institute Press.

Colby, A. et al. eds. 2003. *Educating Citizens: Preparing America's Undergraduates for Lives of Moral and Civic Responsibility*. San Francisco: Jossey-Bass.

Condit, C.M. 1999. Crafting virtue: the rhetorical construction of public morality, in *Contemporary Rhetorical Theory: A Reader*, edited by J.L. Lucaites, C.M. Condit, and S. Caudill. New York: The Guilford Press, 306-25.

Connable, B. 2009. All our eggs in a broken basket: how the Human Terrain System is undermining sustainable military cultural competence. *Military Review* [Online], March-April. Available at: http://usacac.army.mil/CAC2/MilitaryReview/Archives/English/MilitaryReview_20090430_art010.pdf [accessed: January 15, 2010].

Connors, R.J. 1997. *Composition-Rhetoric: Backgrounds, Theory and Pedagogy*. Pittsburgh: University of Pittsburgh Press.

Corbett, E.P.J. 1989. Rhetoric, the enabling discipline, in *Selected Essays of Edward P.J. Corbett*, edited by R.J. Connors. Dallas: Southern Methodist University Press, 194-208.

Corley, W.C. 2006. An officer and a professor. *The Chronicle of Higher Education,* March 24, B14.

Cowan, J.S. 2001. RMC and the profession of arms: looking ahead at Canada's military university. *Canadian Military Journal* [Online], 2(3). Available at: http://www.journal.dnd.ca/ [accessed: November 1, 2009].

Crawford, J. 2006. *The Last True Story I'll Ever Tell*. New York: Riverhead Books.

Edelman, L. 1994. *Homographesis: Essays in Gay Literary and Cultural Theory*. New York: Routledge.

Farquhar, S. 2008. *Interview with Lieutenant Colonel Scott Farquhar*. [Online: Operational Leadership Experiences in the Global War on Terrorism]. Available at: http://carl.army.mil/contentdm/home.htm [accessed: January 27, 2010].

Farrell, A. 2004. Two cultures in military education. *Academic Exchange Quarterly*, 8(2), 173-78.

Felter, J. and Fishman, B. 2007. *Al Qaeda's Foreign Fighters in Iraq: A First Look at the Sinjar Records*. [Online, December 19]. Available at: http://ctc.usma.edu/harmony/Harmony_menu.asp [accessed: January 27, 2010].

Fish, L. 1989. General Edward G. Lansdale and the folksongs of Americans in the Vietnam War. *Journal of American Folklore*, 102(406), 390-411.

Fish, S. 2003. Aim low. *The Chronicle of Higher Education*, May 16, C5.

Fitch, N.R. 1999. *Appetite for Life*. New York: Anchor Books.

Fleming, B. 2005. *Annapolis Autumn: Life, Death, and Literature at the US Naval Academy*. New York: The New Press.

Fontenot, G. 2008. The US Army and contemporary military history. *Army History*, Spring, 26-31.

Ford, J.J. 2005. *Bonk on the Head*. Roberts Creek, BC: Nightwood Editions.

Gates, R.M. 2008. Beyond guns and steel: reviving the nonmilitary instruments of American power. *Military Review*, January-February, 2-9.

Gentile, G.P. 2009. The selective use of history in the development of American counterinsurgency doctrine. *Army History*, Summer, 21-35.

George, A. 2001. Critical pedagogy: dreaming of democracy, in *A Guide to Composition Pedagogies*, edited by G. Tate, A. Rupiper, and K. Schick. New York: Oxford University Press, 92-112.

Glanz, J. 2007. Death toll in Iraq bombings rises to 250. *The New York Times*, August 15.

Glenn, D. 2009. Program to embed anthropologists with military lacks ethical standards, report says. *The Chronicle of Higher Education* [Online], December 3. Available at: http://chronicle.com/article/Program-to-Embed-Anthropolo/49344/ [accessed: December 4, 2009].

Glickman, C.D. 2002. A dollar could have bought a lot more. *Columns*, University of Georgia News Service, January 28, 7.

Gonzalez, J.J. 2008. How good scholarship makes good citizens. *The Chronicle of Higher Education*, September 19, A34.

Gonzalez, R.J. 2007. We must fight the militarization of anthropology. *The Chronicle of Higher Education*, February 2, B20.

Hammond, W.M. 2008. "The US Army and contemporary military history": some further considerations. *Army History*, Fall, 18-24.

Hanley, B. 2008. Why study the literature of war? *War, Literature, and the Arts*, 20(1-2), 15-18.

Hanley, B. 2009. The Navy is giving history a bad name. *United States Naval Institute Proceedings*, 135, 10.

Hanley, J. 2006. *The German Prisoner*. Holstein, ON: Exile Editions.

Haycock, R.G. 2001. The labours of Athena and the Muses: historical and contemporary aspects of Canadian military education. *Canadian Military Journal* [Online], 2(2). Available at: http://www.journal.dnd.ca/ [accessed: November 1, 2009].

Hemingway, E. 1997. *A Farewell to Arms*. New York: Scribner Classics Edition.

Hodges, P. 2008. "They don't like it up 'em!": bayonet fetishization in the British Army during the First World War. *Journal of War and Culture Studies*, 1(2), 123-38.

Holder, L.D. and Murray, W. 1998. Prospects for military education. *Joint Force Quarterly*, Spring, 81-90.

Houle, C.O. 1961. *The Inquiring Mind.* Madison, WI: University of Wisconsin Press.

Howard, H. 2006. Love, that doth reign and live within my thought, in *The Norton Anthology of English Literature: The Major Authors*, 8th edition, Vol. A, edited by S. Greenblatt et al. New York: Norton, 354.

Howard, M. 1962. The use and abuse of military history. *RUSI Journal*, 107, 4-11.

Howard, M. 2006. Military history and the history of war, in *The Past as Prologue: The Importance of History to the Military Profession*, edited by W. Murray and R.H. Sinnreich. Cambridge, MA: Cambridge University Press, 12-20.

Hwang, D.H. 1988. *M. Butterfly*. New York: Plume.

Isocrates. 1928-1945. *Isocrates*. 3 vols. Translated by G. Norlin and L. Van Hook. Cambridge, MA: Harvard University Press.

Ivorybill. 2007. *Comments posted in response to "Informed Comment: Thoughts on the Middle East, History, and Religion."* [Online, August 16]. Available at: http://www.juancole.com/2007/08/thousands-of-sadrists-protest-sinjar.html [accessed: January 27].

Jacoby, B. 2009a. Civic engagement in today's higher education: an overview, in *Civic Engagement in Higher Education: Concepts and Practices*, edited by B. Jacoby and Associates. San Francisco: John Wiley and Sons, 5-30.

Jacoby, B. 2009b. Preface. *Civic Engagement in Higher Education: Concepts and Practices*, edited by B. Jacoby and Associates. San Francisco: John Wiley and Sons, ix-xii.

Johnson, B.W. and Harper, G.P. 2005. *Becoming a Leader the Annapolis Way*. New York: McGraw-Hill.

Katz, A. 1995. Pedagogy, resistance, and critique in the composition class, in *Left Margins: Cultural Studies and Composition Pedagogy*, edited by K. Fitts and A.W. France. Albany, NY: State University of New York Press, 209-18.

Kidwell, D.C. 2005. *Public War, Private Fight?: The United States and Private Military Companies*. Fort Leavenworth, KS: Combat Studies Institute Press.

Kohn, R., ed. 1997. The practice of military history in the US government: the Department of Defense. *Journal of Military History*, 61, 121-47.

Kohn, R. 2001. *Soldiers and Civilians: The Civil-Military Gap and American National Security*. Cambridge, MA: Massachusetts Institute of Technology Press.

Kohn, R. 2008. Coming soon: a crisis in civil-military relations. *World Affairs*, Winter, 69-80.

Kroll, A. 2009. *The Military-Corporate Legacy of the New Secretary of Education*. [Online, January 18]. Available at: http://www.tomdispatch.com/post/175022/andy_kroll_will_public_education_be_militarized [accessed: January 20, 2009].

Kureishi, H. 2006. My Beautiful Launderette, in *The Longman Anthology of British Literature, Vol.2C*, third edition, edited by K.J.H. Dettmar and J. Wicke. New York: Pearson Longman, 2942-86.

Kusiak, P. 2008. Sociocultural expertise and the military: beyond the controversy. *Military Review*, November-December, 65-76.

Lansdale, E. 1991. *In the Midst of Wars: An American's Mission to Southeast Asia.* New York: Fordham University Press.

Leggatt, A. 2005. *Shakespeare's Tragedies: Violation and Identity*. Cambridge: Cambridge University Press.

Lévy, B.-H. 2008. *Left in Dark Times: A Stand against the New Barbarism.* New York: Random House.

Little, G. 2007. *Interview with Sergeant First Class Gary Little*. [Online: Operational Leadership Experiences in the Global War on Terrorism]. Available at: http://carl.army. mil/contentdm/home.htm [accessed: January 27, 2010].

Longo, N.V. and Shaffer, M.S. 2009. Leadership education and the revitalization of public life, in *Civic Engagement in Higher Education: Concepts and Practices*, edited by B. Jacoby and Associates. San Francisco: John Wiley and Sons, 154-73.

Loveland, A.C. 1997. *American Evangelicals and the US Military, 1942-1993*. Baton Rouge, LA: Louisiana State University Press.

Lucaites, J. and Condit, C. 1999. Introduction, in *Contemporary Rhetorical Theory: A Reader*, edited by J.L. Lucaites, C.M. Condit, and S. Caudill. New York: The Guilford Press, 1-18.

Luvaas, J. 1982. Military history: is it still practicable? *Parameters*, March, 2-24.

Lynn, J.A. 2008. Breaching the walls of academe: the purposes, problems, and prospects of military history. *Academic Questions*, 21, 18-36.

Maisel, S. 2008. Social change amidst terror and discrimination: Yezidis in the new Iraq. *The Middle East Institute Policy Brief* [Online], 18 (August). Available at: http://www.mei.edu/Portals/0/Publications/Yezidis-in-the-New-Iraq.pdf [accessed: January 27, 2010].

Marine Corps Intelligence Activity. 2005. *Iraq Culture Smart Card.* [Online]. Available at: http://cryptome.org/iraq-culture.htm [accessed: January 24, 2010].

Matthews, L.J. 2002a. The uniformed intellectual and his place in American arms, part I. *Army*, 52, 17-25.

Matthews, L.J. 2002b. The uniformed intellectual and his place in American arms, part II. *Army*, 52, 31-40.

McComiskey, B. 2000. *Teaching Composition as a Social Process*. Logan, UT: Utah State University Press.

McFate, M. 2005. Anthropology and counterinsurgency: the strange story of their curious relationship. *Military Review*, March-April, 24-38.

McShane, T.W. 2008. A look at "CGSC South" Fort Gordon's satellite campus for ILE. *Command and General Staff College Foundation News*, 4 (Spring), 28-9.

Meier, K. 2006. *Interview with First Lieutenant Karl Meier*. [Online: Operational Leadership Experiences in the Global War on Terrorism]. Available at: http://carl.army.mil/contentdm/home.htm [accessed: January 27, 2010].

Mershon, S. and Schlossman, S.L. 1998. *Foxholes and Color Lines: Desegregating the US Armed Forces*. Baltimore: John Hopkins University Press.

Miller, T.P. 1997. *The Formation of College English: Rhetoric and Belles Lettres in the British Cultural Provinces*. Pittsburgh: University of Pittsburgh Press.

Millett, A.R. 1992. American military history: Clio and Mars as "Pards," in *Military History and the Military Profession*, edited by D.A. Charters, M. Milner, and J.B. Wilson. Westport, CT: Praeger, 3-21.

Moore, B. 1993. *A Lexicon of Cadet Language: Royal Military College, Duntroon, in the Period 1983-1985*. Canberra, Australia: Australian National University Press.

Mouffe, C. 1991. Democratic citizenship and political community, in *Community at Loose Ends*, edited by Miami Theory Collective. Minneapolis, MN: University of Minnesota Press, 70-82.

Niday II, J.A. and Harrington, K. 2007. Can academic freedom work in military academies? *Academe*, 93(4), 26-29.

Norlin, G. 1928. *Isocrates*. Vol. 1. Cambridge, MA: Harvard University Press.

Obama, B. 2009. *Remarks by the President at the United States Naval Academy Commencement*. [Online, May 22]. Available at: http://www.whitehouse.gov/the_press_office/Remarks-by-the-President-at-US-Naval-Academy-Commencement [accessed: June 22, 2009].

Ohmann, R. 1976. *English in America: A Radical View of the Profession*. New York: Oxford University Press.

Olson, C.T. 2007. *So This is War: A 3rd US Cavalry Officer's Memoirs of the Triumphs, Sorrows, Laughter, and Tears During a Year in Iraq*. Bloomington, Indiana: Authorhouse.

Palm Center. 2006. *Poll: Majority of Troops Would Welcome Gay Soldiers*. [Online: Palm Center: Blueprints for Sound Public Policy]. Available at: http://www.palmcenter.org/press/dadt/in_print/poll_majority_of_troops_would_welcome_gay_soldiers [accessed: June 1, 2009].

Palm, E.F. 1985a. Book burning at Berkeley. *San Francisco Chronicle*, March 2, 34.

Palm, E.F. 1985b. ROTC doesn't promote militarism. *The Daily Californian*, October 10, 4.

Palm, E.F. 1986a. Winning hearts and minds at Berkeley. *Marine Corps Gazette*, February, 61-5.

Palm, E.F. 1986b. Novels of Vietnam and the uses of war literature. *Marine Corps Gazette*, November, 92-9.

Palm, E.F. 1996. Putting the Naval Academy on course. *The Chronicle of Higher Education*, May 10, A60.

Palm, E.F. 1998. Real officers don't teach Keats: what's really rotten in Annapolis. *Roofbeam*, 1(1), 22-3, 29-31.

Palm, E.F. 2008. Women warriors: shouldn't we talk about this? *The Seattle Times*, January 18.

Perry, W.G., Jr. 1963. Examsmanship in the liberal arts: a study in educational epistemology. *Examining in Harvard College: A Collection of Essays by Members of the Harvard Faculty.* Available at: http://www.people.fas.harvard.edu/~lipoff/miscellaneous/exams.html [accessed: June 1, 2009].

Peters, R. 2007. Learning to lose. *The American Interest* [Online], July-August. Available at: http://www.the-american-interest.com/article.cfm?piece=291 [accessed: January 20, 2009].

Petraeus, D.H. 2007. Beyond the cloister. *The American Interest* [Online], July-August. Available at: http://www.the-american-interest.com/article.cfm?piece=290 [accessed: January 20, 2009].

Poulakos, J. 1989. Early changes in rhetorical practice and understanding: from the Sophists to Isocrates. *Texte*, 8(9), 307-24.

Poulakos, T. 1997. *Speaking for the Polis: Isocrates' Rhetorical Education.* Columbia, SC: University of South Carolina Press.

Prep-USA. 2005. *PREP Mission Statement* [Online, February 15]. Available at: http://www.prep-usa.org/portal/main/generaldetail.asp?ID=296[accessed:June 26, 2009].

Price, D. 2008. *Anthropological Intelligence: The Deployment and Neglect of American Anthropology in the Second World War*. Durham, NC: Duke University Press.

Reitan, J. 2005. Report on the Equality Ride's Journey to the United States Naval Academy. [Online: Soulforce]. Available at: http://www.soulforce.org/article/720 [accessed: June 2, 2009].

Remarque, E.M. 1996. *All Quiet on the Western Front*. Translated by B. Murdoch. London: Vintage Books.

Report of the Special Committee to the Board of Visitors. 1997. *The Higher Standard: Assessing the United States Naval Academy*. United States Naval Academy.

Rice, A. 2002. "A Peculiar power about rottenness": annihilating desire in James Hanley's *The German Prisoner*. *Modernism/modernity*, 9(1), 75-89.

Ricks, T.E. 1997. The widening gap between the military and society. *Atlantic Monthly*, July, 66-78.

Ricks, T.E. 2009. Why we should get rid of West Point. *The Washington Post*, April 19.

Robertson, C. and Farrell, S. 2009. Ambush by an ally chills trust in Iraqi units. *The New York Times*, May 7.

Roosevelt. T. 1910. Citizenship in a republic. Speech at the Sorbonne.

Samet, E. 2007. *Soldier's Heart: Reading Literature through Peace and War at West Point*. New York: Farrar, Straus and Giroux.

Sartre, J.-P. 1989. *No Exit and Three Other Plays*. New York: Vintage International.

Schaeffer, F. 2009. Ivies and the military could learn a lot from each other. *The Boston Globe* [Online, April 9]. Available at: http://www.boston.com/bostonglobe/editorial _opinion/oped/articles/2009/04/09/ivies_and_military_could_learn_a_lot_from_each_othe r/ [accessed: April 10, 2009].

Schiappa, E. 1995. Isocrates' philosophia and contemporary pragmatism, in *Rhetoric, Sophistry, Pragmatism*, edited by S. Mailloux. New York: Cambridge University Press, 33-60.

Sedgwick, E.K. 1985. *Between Men: English Literature and Male Homosocial Desire*. New York: Columbia University Press.

Shakespeare, W. 2006. *Hamlet*. Edited by A. Thompson and N. Taylor. London: Arden Shakespeare.

Shalikashvili, J. 2007. Second thoughts on gays in the military. *The New York Times* [Online, January 2]. Available at: http://www.nytimes.com/2007/01/02/opinion/02shalikashvili.html?_r=2 [accessed: June 1, 2009].

Shipman, D. 2006. *Interview with Colonel Doug Shipman*. [Online: Operational Leadership Experiences in the Global War on Terrorism]. Available at: http://carl.army.mil/contentdm/home.htm [accessed: January 27, 2010].

Sinfield, A. 1994. *The Wilde Century: Effeminacy, Oscar Wilde, and the Queer Movement*. New York: Columbia University Press.

Sinnreich, R.H. 2006. Awkward partners: military history and American military education, in *The Past as Prologue: The Importance of History to the Military Profession*, edited by W. Murray and R.H. Sinnreich. Cambridge, MA: Cambridge University Press, 55-77.

Spiller, R. 2006. Military history and its fictions. *Journal of Military History*, 70, 1081-97.

Stewart, R.W. 2009. The relevance of military history. *Army History*, Spring, 58-9.

Stiehm, J.H. 2002. *The US Army War College: Military Education in a Democracy*. Philadelphia: Temple University Press.

Stotsky, S. 1992. Conceptualizing writing as moral and civic thinking. *College English*, 54(7),794-808.

Strauss, B. 2008. Military education: models from antiquity. *Academic Questions*, 21, 52- 61.

Taber, N. 2005. Learning how to be a woman in the Canadian Forces/unlearning it through feminism: an autoethnography of my learning journey. *Studies in Continuing Education*, 27(3), 289-301.

"The Cadet Leader Development System and the Academic Program's Goals." 2007. *Educating Future Army Officers for a Changing World*. West Point, NY: Office of the Dean, United States Military Academy.

Toledo Blade. 1990. Inquiry seeks ruling on USNA hazing. May 19.

Turner, B. 2005. *Here, Bullet*. Farmington, ME: Alice James Books.

United States Air Force Academy. 2008. *The Officer Development System: Developing Leaders of Character*. [Online, August 11]. Available at: http://www.e-publishing.af.mil/shared/media/epubs/USAFAPAM36-3527.pdf [accessed: January 15, 2010].

United States Army Combined Arms Center. 2009. *Command and General Staff School*. [Online]. Available at: http://usacac.army.mil/cac2/cgsc/cgss/index.asp [accessed: February 3, 2010].

United States Army and Marine Corps. 2007. *Counterinsurgency Field Manual* (Army Manual No 3-24; Marine Corps publication No 3-33.5). Chicago: University of Chicago Press.

United States Department of Defense. 2008. *Speech delivered to the Association of American Universities by Secretary of Defense Robert M. Gates*. [Online]. Available at: http://www.defense.gov/speeches/speech.aspx?speechid=1228 [accessed: March 15, 2010].

United States Naval Academy. 2009. *United Stated Naval Academy*. [Online, May 29]. Available at: http://www.usna.edu///homepage.php [accessed: June 1, 2009].

USS Essex. 2009. *Mission*. [Online] Available at: http://www.essex.navy.mil/Site%20Pages/Ship%20Mission.aspx. [accessed: June 17, 2009].

Vernon, A. 2002. A barbarian in the ivory tower. *The Chronicle of Higher Education*, February 22, B14-15.

Vernon, A. 2006. *Most Succinctly Bred*. Kent, OH: Kent State University Press.

Vickers, B. 1988. *In Defense of Rhetoric*. New York: Oxford University Press.

Virginia Military Institute. 2009. *Administration* [Online]. Available at: http://www.vmi.edu/show.aspx?tid=36563&id=62. [accessed June 15, 2009].

Wander, M. 2006. *Interview with First Lieutenant Mitchell Wander*. [Online: Operational Leadership Experiences in the Global War on Terrorism]. Available at: http://carl.army. mil/contentdm/home.htm [accessed: January 27, 2010].

Webb, J. 1979. Women can't fight. *Washingtonian* [Online], November. Available at: http://washingtonian.com/print/articles/6/173/2182.html [accessed: June 1, 2009].

Webb, J. 1987. *Forrestal Lecture*. [Online]. Available at: http://www.jameswebb.com/speeches/speeches-forrestallecture-usna.htm [accessed: June 1, 2009].

Webb, J. 2006. *Meet the Press Interview*. [Online]. Available at: http://www.youtube.com/watch?v=UebXkJhGpyA [accessed: June 1, 2009].

"West Point Mission." 2009. *Building Capacity to Lead: The West Point System for Leader Development*. West Point, NY: Office of the Dean, United States Military Academy.

Wong, E. 2006a. GI's investigated in slayings of 4 and rape in Iraq. *The New York Times*, July 1.

Wong, E. 2006b. Insurgent group posts video of 2 mutilated US soldiers. *The New York Times*, July 11.

Worth, R.F. 2006. Blast at Shiite shrine sets off sectarian fury in Iraq. *The New York Times*, February 23.

Wright, D.P., and Reese, T.R. 2008. *On Point II: The United States Army in Operation Iraqi Freedom, May 2003-January 2005*. Fort Leavenworth, KS: Combat Studies Institute Press.

Yingling, P. 2006. *Interview with Lieutenant Colonel Paul Yingling*. [Online: Operational Leadership Experiences in the Global War on Terrorism]. Available at: http://carl.army. mil/contentdm/home.htm [accessed: January 27, 2010].

Yingling, P. 2007. A failure in generalship. *Armed Forces Journal* [Online], May. Available at: http://www.armedforcesjournal.com/2007/05/2635198 [accessed: January 10, 2010].

Index

Made in the USA
Monee, IL
13 September 2026

62976738R00116